Document Mechanics:

Basic Tools & Techniques for Variable Content Publishing

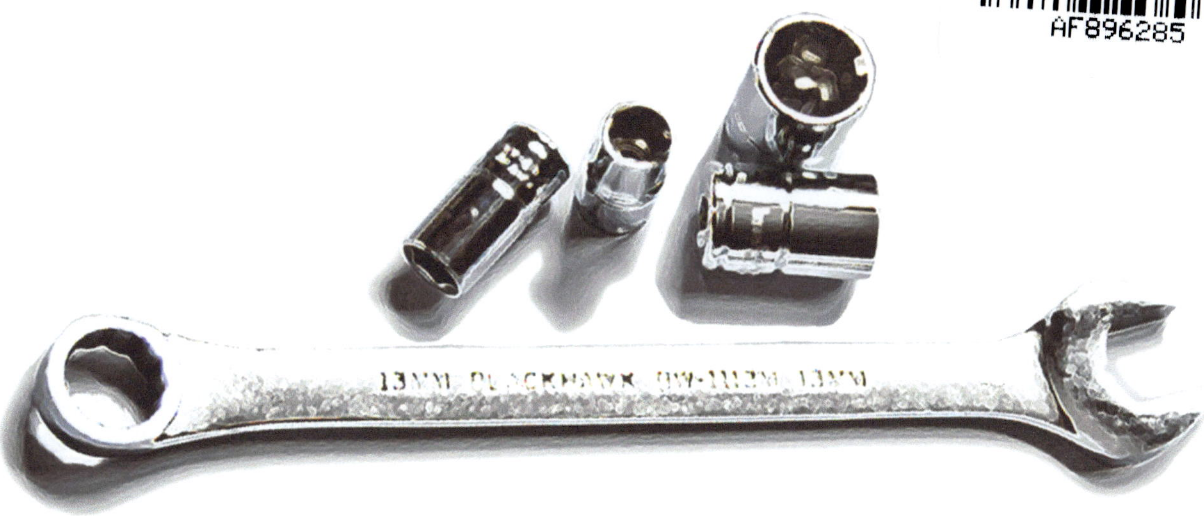

Nicholas D. Barzelay

Document Mechanics: Basic Tools & Techniques for Variable Content Publishing

Published by the author and printed in the United States.

ISBN-13 978-0-615-48281-1

ISBN-10 0-615-48281-3

Content typefaces: Agency FB (headings), Bodoni MT (text), Hypatia Sans Pro (outlines, captions, & XML code), Futura (specifically for Perl program code) and American Typewriter (general use).

Cover typefaces: American Typewriter, Bauhaus 93, Braggadocio, and Lucida Handwriting.

This book was created using Apple iWork Pages, OmniGraffle software by The Omni Group, and Adobe Photoshop, Adobe InDesign, and Adobe Acrobat Professional.

Photographic images used on the cover and throughout the content were captured by the author using a Nikon 5700 set at camera raw resolution. Diagrams used throughout the content were designed by the author and executed using OmniGraffle. All graphic images and diagrams were converted to high resolution JPEG using Photoshop before being placed into the document layout.

Dedication

To my wife, Wendy; my family; and my dogs, Tyler and Maxwell:

for their patience, encouragement, interest, support, and companionship.

This book wouldn't have been possible without them.

Acknowledgements

I'd like to take this opportunity to acknowledge four people whose encouragement, kindness, and help were particularly important in the creation of this book. Each one took time from their very busy schedules to make themselves available for me.

- Dr. Franziska Frey, McGhee Distinguished Professor, School of Print Media, College of Imaging Arts and Sciences at Rochester Institute of Technology
- Patricia Albanese, Gannett Distinguished Professor, School of Print Media, College of Imaging Arts and Sciences at Rochester Institute of Technology
- Nick DiSantis, Manager of Research & Development at Qualprint, A division of Quality Printing Co., Inc. in Pittsfield, Massachusetts
- Kelly J. Mari (Barzelay), my daughter.

Franziska and Pat spent time with me discussing issues, providing opportunities to present subject matter revisions to their graduate and undergraduate classes, and introducing me to other faculty who had related interests. In fact, Franziska originally introduced me to Pat. Through their time and generosity, they helped me stay engaged with R.I.T. while developing portions of this book.

Nick is an alumnus of the Print Media Master's Program at R.I.T. and a fellow classmate. His considerable hands-on knowledge of offset and digital printing was a great help in decisions regarding publishing and distribution. He provided access to a digital press for an initial proof run, which was the basis for subsequent iterations of the physical book layout, resulting in the design you see today.

Kelly took time from her wedding preparations to provide me with digital samples of her wedding invitations and related documents, which were used in Chapter 20. She designed her cards using Adobe Photoshop and Adobe InDesign. At times she worked at the pixel level to insure accuracy and clean lines in her card designs. Her contributions provided the chapter with a live example and thought provoking finish.

Table of Contents

Tables of Figures, Technical Listings, & Checklists x
Figures x
Technical Listings xiii
Checklists xiv

ONE: PERSPECTIVE 1

Chapter 1. Initial Considerations 3
About The Problem 3
What About Training? 4
Is This Realistic? 5

Chapter 2. Patterns 7
Rationale 7
Context 8
Constructs 10
Our Approach 13
Suggested Reading 13

Chapter 3. What Is a Variable Document? 15
Definition 15
Standardization 16
Variation 16
Implications 18
Further Reading 18

TWO: FOUNDATION 19

Chapter 4. Metadata About Digital Assets 21
Overview 21
File Data 22
Camera Data 22
IPTC Core Data 23
Other Data Categories 23
Search, Identification, & Selection 24
Some Explanatory Reading 26

Chapter 5. Metadata About Data 27

Overview 27

Data Files 27

Databases 29

Recommended Reading 33

Chapter 6. XML – Metadata About Text 35

Overview 35

Building Blocks 36

Well Formed 39

Valid 39

Keep In Mind 43

Recommended Reading 43

Supplemental Information 44

THREE: MECHANICS 45

Chapter 7. DAM – Managing Digital Assets 47

Beyond Definition 47

Organization 48

Security 50

Space 52

Working With Images 53

Supplemental Reading 55

Chapter 8. Desktop DBMS – Managing Source Data 57

Data Acquisition 57

Basic Processing Patterns 58

Data Import 62

Data Enhancement 63

Content Stream Export 69

Supplemental Reading 71

Chapter 9. Techniques – Managing XML Content Streams 73

Next Step 73

About the XML Content Stream 74

Using a Text Editor 75

Using XSLT 77

Using Perl Programming 80

Recommended Reading 86

Chapter 10. Document Design – Integration & Print Stream 87
A Rough Sketch 87
Geometry 88
Page Masters 90
Styles 90
Print Stream 91
Recommended Reading 93

FOUR: WORKFLOW 95

Chapter 11. An Iterative VCP Workflow 97
Rationale 97
Good Intentions 98
Depiction 99
Definition 100
Implications 101

Chapter 12. Nothing Flashy – A Basic VCP Tool Kit 105
Rationale 105
Development Platform 105
Development Environment 107
Self-study Suggestions 110

Chapter 13. Intent, Performance, & Quality 111
Practical Intentions 111
Setting Objectives 111
Identifying Requirements 114
Planning for Performance 118
Pursuing Quality 120
Supplementary Reading 124

FIVE: DEMONSTRATIONS 125

Chapter 14. Demonstration Overview 129
Background 129
Approach 130

Chapter 15. A Simple Text Demonstration 131
Purpose....................131
Description....................131
Components....................132
Assembly....................134
Observations....................135

Chapter 16. The "Articles" Demonstration....................................137
Purpose....................137
Description....................137
Components....................138
Assembly....................141
Observations....................143

Chapter 17. The "Editions" Demonstration 145
Purpose....................145
Description....................145
Components....................146
Assembly....................148
Observations....................150

Chapter 18. The "Poems" Demonstration 151
Purpose....................151
Description....................151
Components....................152
Assembly....................153
Observations....................155

Chapter 19. The "Mailer" Demonstration 157
Purpose....................157
Description....................157
Components....................158
Assembly....................160
Observations....................161

Chapter 20. Cards – Analysis, Conjecture, & Possibility 163
Purpose....................163
Alternatives....................163
Foundations....................164
Decisions....................165
Possibilities....................167

Chapter 21. A Publishing Encore **169**
Implication 169
On Reflection 169
Who & Why? 170
Printing Issues 171
Ink on What? 172
Intimation 173

SIX: Information 175

Reference **177**

About the Author **180**

Appendix A. Abbreviations & Acronyms **181**

Appendix B. Perl Content Stream Utilities **183**
Overview 183
About Use 184
Convert 184
Insert 187
Delete 188
Change 189
Reorder 191
Combine 192
Merge 194
Split 195

Table of Figures, Listings, & Checklists

Figures

Figure 2-1. The Demographic Context for VCP 8
Figure 2-2. The Operational Context for VCP 9
Figure 2-3. Business Intelligence Production Activities 10
Figure 2-4. VCP Solution Design Elements and Relationships 11
Figure 2-5. An Approach for Individual and Group Personalization 11
Figure 2-6. An Approach for Incremental Production and Local Personalization 12

Figure 3-1. Document Standardization 16
Figure 3-2. Document Variation 17

Figure 4-1. A Photograph in Adobe Bridge 21
Figure 4-2. Adobe Bridge File Metadata 22
Figure 4-3. Adobe Bridge Camera Metadata 22
Figure 4-4. Adobe Bridge IPTC Core Metadata 23
Figure 4-5. Key Word Grouping in Adobe Bridge 24
Figure 4-6. Adobe Bridge File Filtering 25

Figure 5-1. A Sequential Data File 28
Figure 5-2. Mac OS X File Header Panel 28
Figure 5-3. A Tab-Delimited Sequential Data File 29
Figure 5-4. Relational Data Model 30
Figure 5-5. One-to-many Example 30
Figure 5-6. Database Table Model 31
Figure 5-7. Parent Table Example 32
Figure 5-8. Child Table Example 32

Figure 7-1. Content Choreography – A Dance Overview 47
Figure 7-2. Content Storage Partitioning and Circulation 49
Figure 7-3. Image Preparation Workflow 53

Figure 8-1. Data Acquisition, Preparation, and Storage ... 58
Figure 8-2. Sequential File Processing Model ... 59
Figure 8-3. Sequential File Merge Processing ... 60
Figure 8-4. Database Table Processing Model ... 61
Figure 8-5. Data Preparation Workflow ... 62
Figure 8-6. Script for Inserting Column Values ... 68
Figure 8-7. Script for Selective Value Insertion ... 68
Figure 8-8. Script to Copy and Paste Values ... 69
Figure 8-9. Script to Combine Multiple Values and Then Insert the Result ... 69

Figure 9-1. XML Content Stream Preparation Workflow ... 73
Figure 9-2. XSLT Process Overview ... 77
Figure 9-3. XML DOM Processing Model ... 77
Figure 9-4. XML Sequential Processing Model ... 80
Figure 9-5. Sequential XML Processing Logic ... 81
Figure 9-6. Sequential XML Processing Flow ... 81
Figure 9-7. Basic Program Architecture ... 82

Figure 10-1. Design Geometry (Images, Data, and Text) ... 88
Figure 10-2. Print Stream from Single Page Spreads ... 92
Figure 10-3. Print Stream from Facing Page Spreads ... 92

Figure 11-1. Basic VCP Integration Model ... 98
Figure 11-2. XML VCP System Workflow ... 99
Figure 11-3. XML VCP Interactive Workflow ... 102
Figure 11-4. Workflow Interactivity ... 103

Figure 12-1. Basic System Architecture ... 105
Figure 12-2. Basic VCP Development Environment ... 107

Figure 13-1. Setting Objectives ... 112
Figure 13-2. Traceability in Project Planning & Budgeting ... 114
Figure 13-3. Reflection – Objectives, Digital Assets, & Data ... 117
Figure 13-4. Planning Elements ... 119
Figure 13-5. Quality: Standards, Observations, & Evaluations ... 120

Figure 16-1. The Weekly Article – An Example 137
Figure 16-2. A Common Design Approach 140
Figure 16-3. Weekly Article Designed With a Single Master Text Frame 141
Figure 16-4. Weekly Article XML Reload 142
Figure 16-5. Another XML Reload 142

Figure 17-1. Page Mastesr for First Page (front) and Subsequent Pages (back) 148
Figure 17-2. Monthly Newsletter (Pages 1 to 3) 149

Figure 18-1. The Poems' Page Master 153
Figure 18-2. Example Poem One 154
Figure 18-3. Example Poem Two 155
Figure 18-4. Example Poem Three 155

Figure 19-1. The Advertising Mailer (Variable on One Side) 157
Figure 19-2. The Advertising Mailer (Variable Front & Back) 158
Figure 19-3. Mailer Page Master – Variable on One Side 159
Figure 19-4. Mailer Page Master – Variable Front and Back 160
Figure 19-5. Mailer Sample 1 160
Figure 19-6. Mailer Sample 2 161
Figure 19-7. Mailer Sample 3 161

Figure 20-1. Greeting Card Mockup #1 163
Figure 20-2. Greeting Card Template 164
Figure 20-3. Greeting Card Data Source 164
Figure 20-4. Greeting Card Mockup #2 165
Figure 20-5. Wedding Invitation 167
Figure 20-6. Reception Invitation 167
Figure 20-7. RSVP Card (Front) 168
Figure 20-8. RSVP Card (Back) 168
Figure 20-9. Map & Directions Card (Front) 168
Figure 20-10. Map & Directions Card (Back) 168

Technical Listings

Listing 6-1. Short XML Example ... 38
Listing 6-2. The "Sack Lunch" DTD ... 41
Listing 6-3. The "On-hand Inventory Spreadsheet" DTD ... 41
Listing 6-4. A Variable Document XML Content Stream With Two Records (rows) ... 42
Listing 6-5. DTD for the Variable Document XML Content Stream ... 42

Listing 8-1. Record Layout Before Changes ... 67

Listing 9-1. The Dog Database ... 74
Listing 9-2. The Top plus One Record of an FMPXMLRESULT Export ... 74
Listing 9-3. FMPDSORESULT Export Results (all three records) ... 75
Listing 9-4. Clean XML Content Stream ... 76
Listing 9-5. The Raw XML "Cleanup" Stylesheet ... 78
Listing 9-6. The Image Element "Conversion" Stylesheet ... 79
Listing 9-7. Perl Program Code for XML Content Stream Cleanup ... 83 to 84

Listing 10-1. A Truncated XML Content Stream ... 91

Listing 15-1. A Report DTD ... 132
Listing 15-2. XML-tagged Front Matter ... 133
Listing 15-3. Portion of XML from Report Body ... 133
Listing 15-4. Example XML Bibliographic Citation ... 133
Listing 15-5. Report CSS Segment ... 134

Listing 16-1. Weekly Article DTD ... 138
Listing 16-2. Weekly Article XML ... 139

Listing 17-1. DTD for Monthly Newsletter ... 146
Listing 17-2. Monthly Newsletter XML Content (1 of 3 similar articles) ... 147

Listing 18-1. The Working Poem DTD ... 152
Listing 18-2. An XML-tagged Poem ... 152
Listing 18-3. Initial Poem DTD – Line Detail Impedes Needed Flexibility ... 156

Listing 19-1. Advertising Mailer DTD ... 158
Listing 19-2. Advertising Mailer XML ... 159

Listing B-1. Convert Tab-delimited Files to XML Content Streams .... 185 to 187
Listing B-2. Insert new XML Elements Into the Content Stream .... 187 to 188
Listing B-3. Delete XML Elements From the Content Stream .... 188 to 189
Listing B-4. Change Selected XML Elements in the Content Stream .... 189 to 190
Listing B-5. Reorder XML Child Elements Within Each <row> Set .... 191 to 192
Listing B-6. Combine Contiguous XML Child Elements Within Each <row> Set .... 192 to 194
Listing B-7. Merge (Append) Multiple XML Content Streams 194
Listing B-8. Split a Single XML Content Stream into Multiple Streams .... 195 to 196

Checklists

Checklist 13-1. Outward & Inward Perspectives 112
Checklist 13-2. Motive, Means, & Opportunity 113
Checklist 13-3. Measuring Achievement 113
Checklist 13-4. Design Requirements 115
Checklist 13-5. Digital Asset Requirements 115
Checklist 13-6. Data Requirements 116
Checklist 13-7. Production Requirements 116 to117
Checklist 13-8. Planning Questions for Order of Execution 118
Checklist 13-9. Tracking Status 119

ONE: Perspective

This is the introductory unit for this book. Very simply, the intent is to provide some background, summarize the perceived problem, and establish a starting point for presenting a multi-fasceted approach to assembling variable document solutions.

1. Initial Considerations

This chapter explains the problem the author believes has limited variable document adoption rates as one directly related to processes, tools, and techniques. The chapter goes on to suggest an overview of a solution. It wraps up by briefly examining issues related to providing comprehensive document development training.

a. About The Problem
b. What About Training?
c. Is This Realistic?

2. Patterns

Next step, the environment for VCP (Variable Content Publishing) is briefly viewed in an organizational and operational context. This provides a conceptual picture of communicating and interacting with the consumers of information and the customers of products and services through the use of variable documents.

a. Rationale
b. Context
c. Constructs
d. Our Approach
e. Suggested Reading

3. What Is a Variable Document?

All this talk about variability is predicated on defining the difference between static and variable documents. We discuss the two primary properties of variable documents: variation and standardization We explore their interrelationship, whether variation and standardization are mutually exclusive or codependent.

a. Definition
b. Standardization
c. Variation
d. Implications
e. Further Reading

1. Initial Considerations

Discussion Points:

- ✓ About The Problem
- ✓ What About Training?
- ✓ Is This Realistic?

About The Problem

Printer manufacturers and allied software developers began promoting versions of variable data printing (VDP) for digital printers in the mid to late 1990's. Although its ability to successfully attract customers, grow revenues, and reduce marketing costs has been statistically validated by industry analysts, adoption of this approach remaines limited, particularly among those who could profit most from it. Frank Romano (2010), a respected print industry expert, wrote:

> "VDP volume is less than 10 percent of all digital printing and digital printing is less than 15 percent of all printing. The use of variable data printing has been retarded over the last decade by short-sighted suppliers, specifications, and associations."

VDP, or as we call it, variable content publishing (VCP), is a major reason for adopting digital technology. An interesting result is the continued predominance of mail addressing – an already existing capability in offset printing. So, what's the big deal? The answer is, it's easy to do. The reasons for it being easy to do are familiarity, pre-packaged address files, and VDP application interfaces that seemingly "hide" data architecture and programming issues. A substitute for the word, "hide", could be "simplify". What "hide" or "simplify" may actually mean is to lower technical features, diminish flexibility, or simply "dumb down" applications.

It seems Industry suppliers continue to devise increasingly elaborate, "user-friendly" interfaces to support traditional graphic arts workflows while trying to simplify adopting VDP (VCP). Their concentration on the user interface doesn't address basic processes underpinning the entire workflow. Concentration on the user interface tends to treat variable document development as a single technology based on interacting with an orderly sequence of screens, panels, and buttons intended to produce addressed documents for marketing communications, financial ststements, and charity appeals.

But there is much more that VCP can produce, and many more aspects involved in producing it. Actually, development efforts integrate multiple technologies in a highly dynamic and potentially disordered set of processes. It means that creating variable documents can be complex, messy, time-consuming, and frustrating – hence the exceedingly slow adoption rate for variable documents and digital production. What do we need?

Randy Pausch, then a professor at Carnegie-Mellon University, stated a key principle while delivering "The Last Lecture" (2007): "**You have to know the basics before you can do the fancy stuff.**" The user interface won't "protect" the document developer from the needs to understand basic processes – how individual components are created and then brought together.

What we need to do is to simplify. Simplifying does not refer to document designs and production output (all too frequent misunderstandings), but rather applies to removing complexity from the workflow. Well, how does a person do that?

1. Identify fundamental workflow processes
2. Identify the most rudimentary possible toolset that can adequately support the processes.
3. Identify a set of rudimentary technical skills so a developer is able to understand basic requirements and therefore know how to apply the tools.

Processes and tools need to be easy to understand (learn) and easy to connect together (integrate). By simplifying, a person may ultimately create a greater variety of increasingly complex variable products than what they could create with more complex and expensive tools based on a graphical user interface.

And, by the way, let's not limit our considerations to just putting ink on paper. A document doesn't have to be printed on a tangible substrate. That document could just as easily be "printed" to the Web, or put in digital format for display on some computing or communication device. When you read "variable" document, keep in mind that it may also be a "virtual" document.

What About Training?

After identifying basics, the issue becomes education and training. Vendor-supplied training for designing and developing variable documents may not provide adequate exposure to fundamental technologies, processes, or underlying concepts. Time is spent navigating graphical user interfaces or pushing buttons in the proper sequence, excluding key activities beyond the interface.

Degree programs may rarely address the cross-disciplinary nature of the variable document workflow, if they address variable document publishing at all. There may be course requirements across departments, but course integration across the workflow isn't provided; few adequately connects the dots.

Graphic Design and Applied Information Technology programs may not offer any course work about variable documents, although both are intimately part of the document creation process. A major in Applied Information Technology probably won't receive formal training in document layout and design. Conversely, a Graphic Design major isn't apt to receive training in systems analysis, information architecture, programming, or database applications.

The combinations of tools and processes used to produce variable documents form a technology-rich publishing system of considerable application breadth and depth. Breadth refers to the variety of different communication products this system can produce. Depth alludes to the degree of variety and intricacy possible within each type of communication product.

Performing consistent, quality work across the variable content workflow requires an understanding of basic concepts, processes, and tools for doing the essential tasks – handling data, text, and images, and designing document layouts that can accommodate automated content. It requires cross-disciplinary knowledge of business, imaging, graphic design, information processing, and production systems across the entire document development workflow. Below are four questions that may have crossed your mind:

1. How am I going to acquire all the knowledge and skills needed to do this kind of work?
2. Where am I going to find a collection of basic information and guidance in one place?
3. How am I going to sort through available literature, tools, and processes to find a combination that works?
4. How long is it going to take me to become adequately proficient with the workflow, selected tools, and related techniques?

In response, my intention here is to provide a single source that examines basic concepts, technologies, and techniques for each step in creating a variable document. The idea is to simplify the whole development process, starting with basic concepts, adding simple tools and skills, and providing assistance along the way:

- Process diagrams
- Screen captures from relevant application software
- Example documents in various stages of development
- Tested program code and accompanying data and graphic content
- Process checklists
- Explanatory text with embedded examples

This approach will spare you from having to understand everything about software applications and programming to be successful – only a small subset of features needed to create a considerable array of variable documents. These basic features provide an introduction to the interplay of technical specialities and a starting point for further exploration, experimentation, and development. There's more than enough here to get started.

Is This Realistic?

Being "realistic" is a question you will have to answer as the reader. I undoubtedly think so as author, or I wouldn't have written all this in the first place. Here are some factors you may want to consider before heading out:

1. **The material presented here was derived from research and lectures:**
   It is based on research and classroom presentations given at Rochester Institute of Technology, School of Print Media, for classes on digital asset management and cross-media workflow at graduate and undergraduate levels.
2. **The material provides a comprehensive workflow treatment:**
   It presents a full pre-media workflow (project inception to print stream preparation), including requirements planning, project management, and quality assurance – topics needed to manage the workflow.
3. **The set of basic technologies has been limited:**
   Focus is on a very basic set of technologies, featuring Adobe Bridge, Filemaker Pro, XML, Perl programming, and Adobe InDesign (only what is needed to produce a variety of variable documents from simple to more complex).
4. **The level of technical detail has been simplified:**
   Discussion is limited to the features of each technology needed for specific workflow steps. The amount of material has been reduced to the relevant essentials – that makes learning easier.
5. **The material is structured for diverse readers with a variety of uses:**
   It will appeal to professionals, practitioners, teachers, and students in Information Technology, Graphic Design, Marketing, Print Media, Multi-media, as well as manufacturers. It can be used as a text book, reference book, training material, self-study guide, or idea source book.

With this as introduction, lets get started. Like the four adventurers in the "Wizard of Oz", we'll start at the very beginning, stay on the yellow brick road, and avoid the tall grass.

2. Patterns

Discussion Points:

- ✓ Rationale
- ✓ Context
- ✓ Constructs
- ✓ Our Approach
- ✓ Suggested Reading

Rationale

Before delving into the details of variable content printing and publishing (VCP or VDP), let's consider a single question: "Why would we want to use this technology?" Quick answers might include connecting with customers, dealing with high competition, economic turmoil, environmental sustainability, or being on the leading-edge of technology. Such responses are common and predictable.

On second thought, how about a simple response such as taking advantage of good business and communication practices? This means getting the most for your money – cost efficiency and cost effectiveness. Consider this anecdote:

Over a year, I get all manner of communications through the U.S. Mail. My mailbox is stuffed! I may read only 20% of it at best – some isn't even opened. So 80% of the printed communications sent to my home either go into recycling or the landfill. Put another way, 80% of the money spent to tell me something, show me something, or convince me about something is wasted. Now multiply that amount of wasted money by all those other people who do the same thing with their mail that I do with mine. Than's a lot of money!

And by the way, for those taking the digital approach (email, cell phones, et cetera), electronic communications still incur time, resources, energy, and monetary costs. Meanwhile, spam filters, junk mail folders, and digital trash cans work well and are improving all the time. If you're like me, a good precentage of my digital communications (text, voice, video, or otherwise) still find their way to the trash.

Unsolicitated or not, mass communications tend to incur waste while failing to convey the message. It's bad business and poor communicating! To top it off, it certainly doesn't do much to counter high competition and economic turmoil, help foster environmental sustainability, let alone effectively using available technology. Those communications, especially if they are repetitive, may do nothing more than irritate

To do better, we need to understand our social and economic environment – the public and private organizations that touch everyones lives. Then, we need to tailor how we communicate within that environment to be effective. What our focus will be in this book is a set of basic concepts and mechanics for effectively communicating in print as our case in point.

Context

Two books have really influenced my thinking on this subject. I read them at different times. One was published in 2006, the other in 2008. I didn't plan on putting them together – they're actually on different subjects. It was an instant subconscious realization that one was pertinent to the other. The first is Seth Godin's "Small Is the New Big" (2006); the second is Bill Bishop's "The Big Sort: The Clustering of Like-minded America Is Tearing Us Apart" (2008).

Demographics

Bishop (2008) describes a demographic tapestry in considerable flux, fragmenting across the national fabric while coalescing locally at or below the county level. Local amalgamations consequentially present considerable political and market implications.

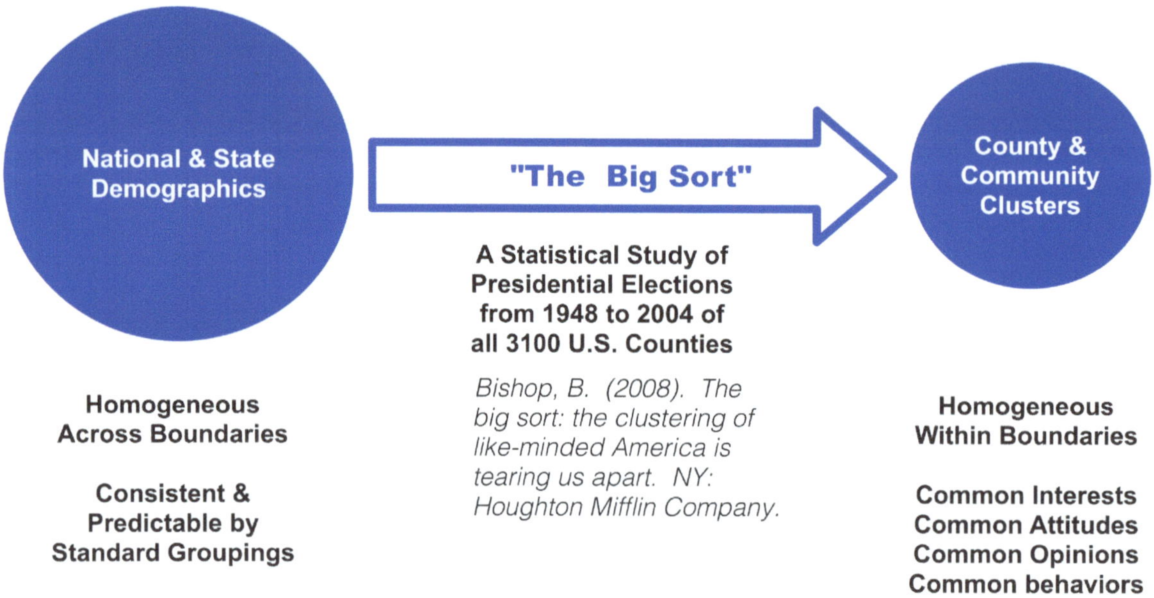

Figure 2-1. The Demographic Context for VCP

National demographics have largely been consistent and predictable by standard groupings – homogeneous across boundaries. The demographic picture is increasingly becoming small clusters of common interests, attitudes, opinions, and behaviors – homogeneous within boundaries (Bishop, 2008). Figure 2-1 illustrates Bishop's idea.

Operations

Godin (2006) discusses the problems of "big" organizations and the advantages of organizations that are "small" or act as if they were small. Small businesses and non-profits tend to be community-centric. Their relationship and communications with customers are (or can be) more personal, informal, face-to-face, and in the form of a dialogue.

Large businesses and non-profits are national and international. Unless they try acting like a small organization, they relate to their customers on a remote, formal, and impersonal level. Their communications (including the telephone with its layers of menu options) are apt to be a frustrating monologue – the big organization telling the little customer how things are. Figure 2-2 below encapsulates Godin's thinking about changing behavioral tendencies as organizations grow or shrink.

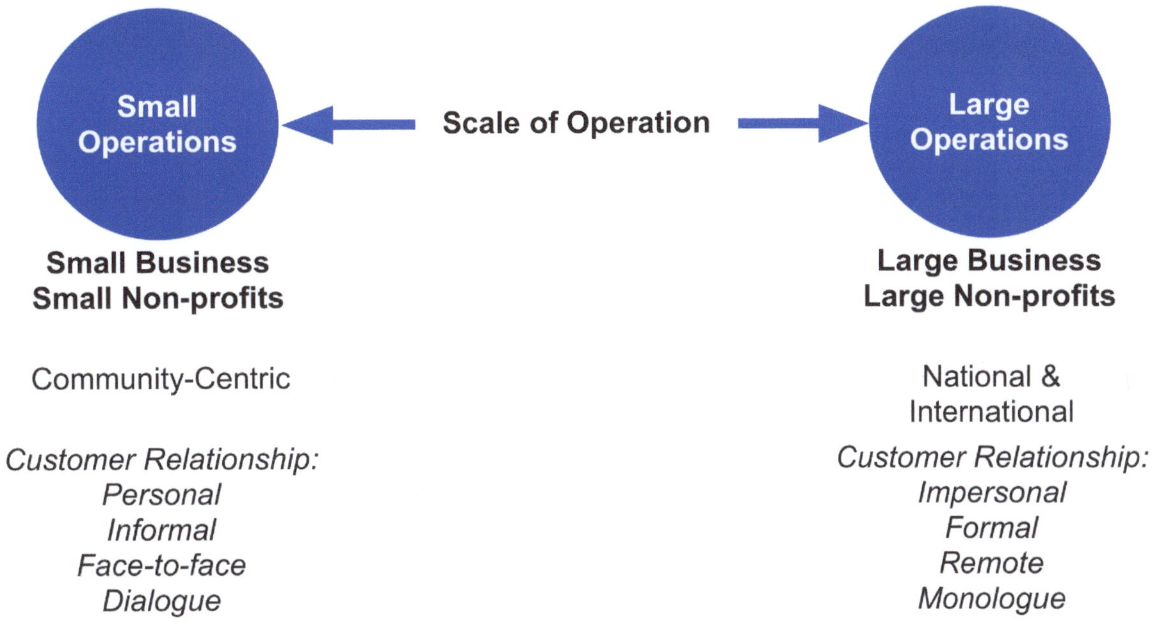

Figure 2-2. The Operational Context for VCP

When Godin's organizational behavioral tendencies are viewed in conjunction with Bishop's demographic findings, we conclude that being small is going to be (or already is) an advantage in building customer relationships. Being big is (or can be) detrimental. One problem is that many small organizations try emulating large ones, and large organizations get larger, act larger, and seem more remote (a computer voice and phone menu, or a voice from off shore replaces human interaction).

Value Proposition

So what does all this have to do with VCP and its digital assets? Just this: VCP provides a communication capability to address individuals or selected groups with information tailored to their interests, attitudes, and behaviors – in other words, targeting customers with the same factors that, according to Bishop (2008), are driving demographic amalgamation. Further, it means communicating with customers in a way similar to a corner hardware or grocery store (Godin's "small"), as opposed to the formal, impersonal, one-way mass marketing campaigns of national chain stores. Another problem is that neither the corner store nor the chain store is likely to exploit the "small" opportunity.

The implication is that being closer to customers improves the chances of meeting organizational objectives (increasing sales, enlarging donations, or improving service relevance). The degree of success directly hinges on how well an organization can develop, acquire, and use customer-related operational intelligence. Figure 2-3 (next page) shows activities that promote this value proposition.

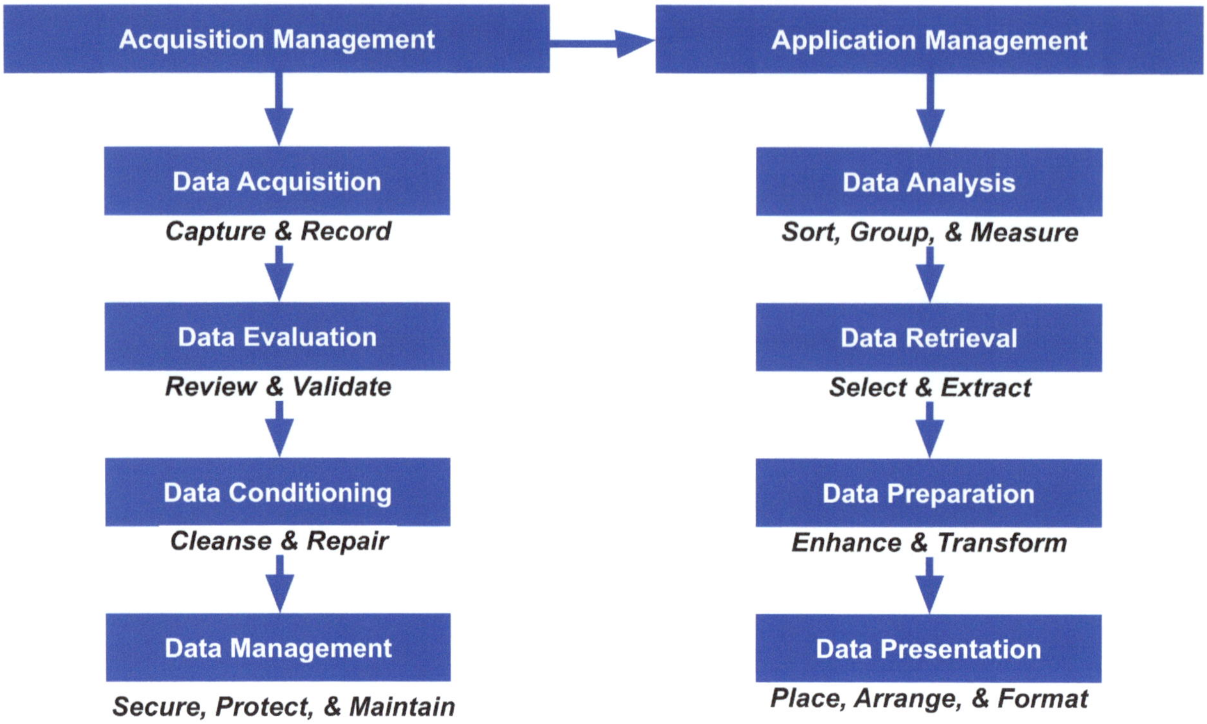

Figure 2-3. Business Intelligence Production Activities

Constructs

Solution Design

The collecting of intelligence makes the rather dubious assumption that ultimately it will be effectively interpreted and intelligently applied. For this discussion, the application is VCP. Data analysis influences both solution design and application. What does a VCP solution design look like?

Figure 2-4 on the next page depicts it. The solution design is dependent on the attributes of four elements: the content source, the production platform, the application purpose, and the desired end products or results. These are the solution requirements.

The design responds to the requirements by assembling system of integrated processes, components, and attributes. The glue between the overall design and each of its elements is a set of relationships: use ("uses a..."), ownership ("owns a..."), or inclusion ("has a..."). The assembled solution design resolves to a system application capable of producing desired end products and results.

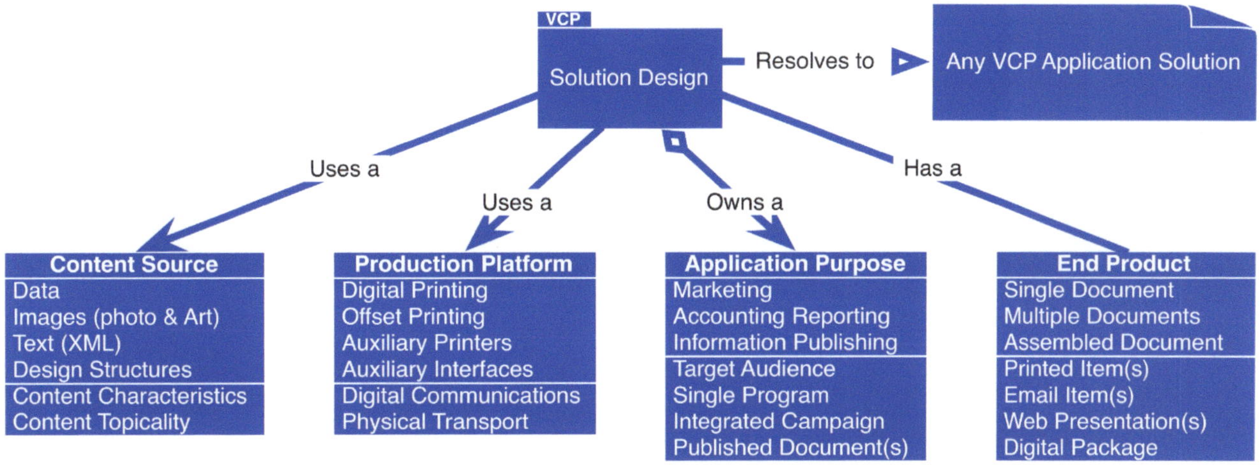

Figure 2-4. VCP Solution Design Elements and Relationships

Now we have a conceptual blueprint to design communication solutions for clients within changing demographic contexts based on relevant operational scale (small or big). As a small organization, we need to take advantage of being small rather than playing at being large. If large, we need to de-emphasizes behaviors associated with "bigness" to appear small ("local"). Keep in mind that we are not necessarily discussing just printed and published documents. We may include documents published on the Internet. This view may also apply to other forms of marketing and sales.

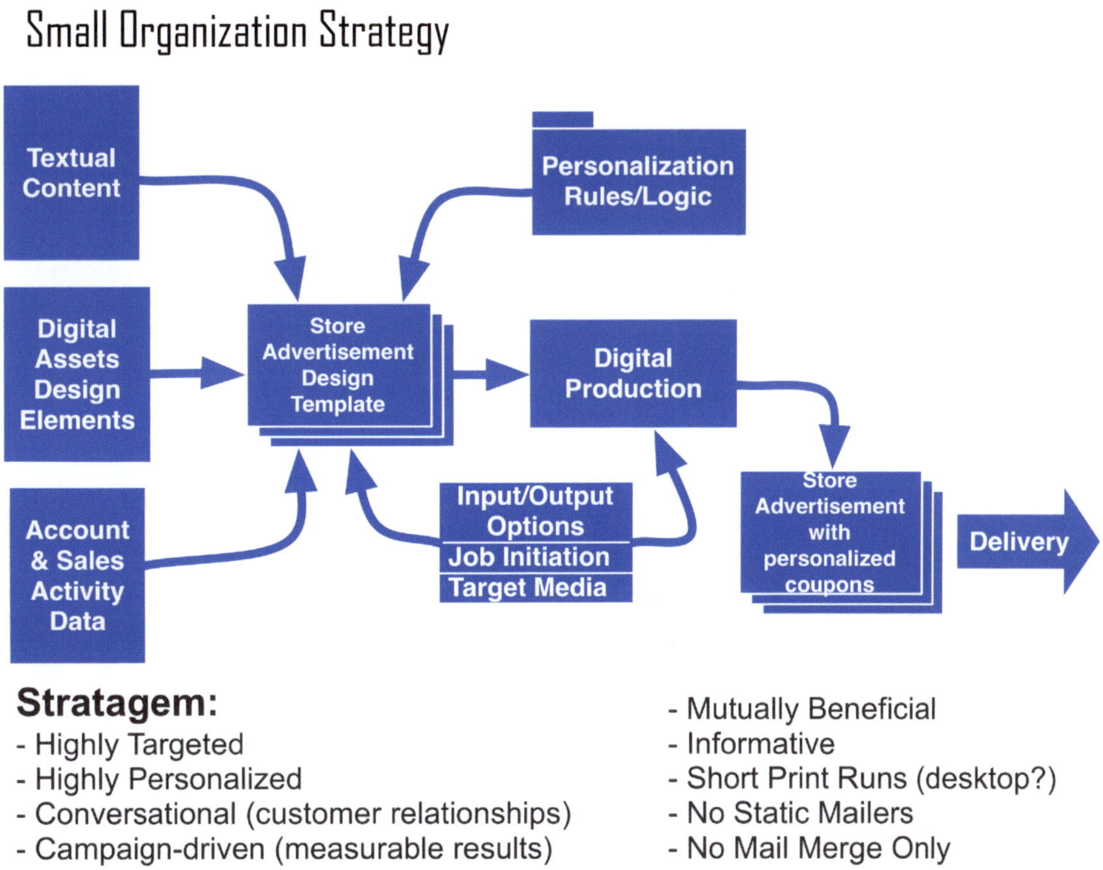

Figure 2-5. An Approach for Individual and Group Personalization

Optimization for this strategy is dependent on how well the small organization records and tracks customer or client interaction. This solution, as shown in Figure 2-5 is not complex. It doesn't have to be! It provides the design basis for any number of custom communication solutions.

Depending on the degree of customization or personalization desired, communications can be targeted to single recipients or to groups (recipients having some common attribute such as gender, interest, location, or something else – perhaps a combination). The key distinction is that selected attributes describe segments of the local population, and not state or national demographic groupings.

Large Organization Strategy

This strategy is more complex because it it has two requirements to fulfill. The first is to satisfy or adhere to organizational branding. The second is to emulate a small organization and address factors (behaviors, attitudes, and preferences for example) descriptive of the local population. Figure 2-6 illustrates this stepped approach.

In the first step, the parent organization creates the basic design, reflecting branding and high level organizational objectives. Most of the design is primarily concerned with static content and some dynamic elements that can be applied at the local level in the second step. The results of first step activity are sent to respective branch locations for local customization and personalization.

The second step applies the dynamic content according to local data. The key to this strategy is the ability to accurately identify, accumulate, analyze, and apply the local data. Since this strategy implies decentralization of printing, a comparative cost-benefit analysis that looks at relative scales of economy to levels of efficacy in achieving objectives may be appropriate.

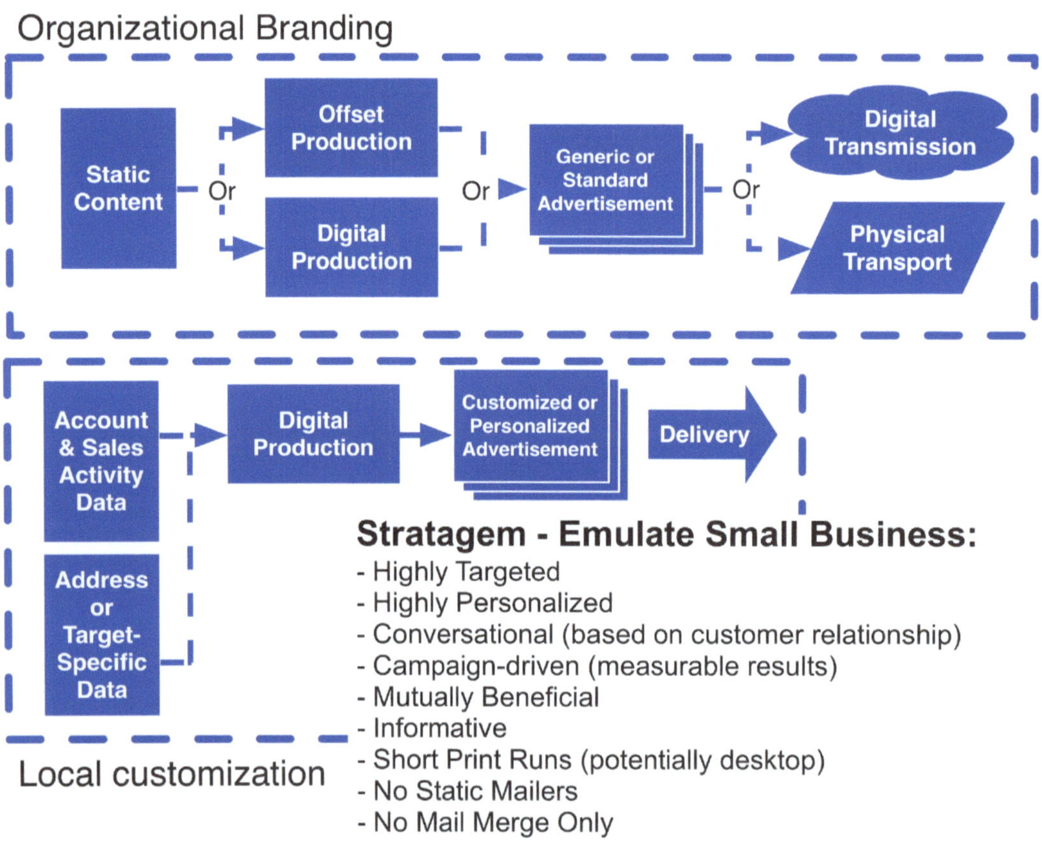

Figure 2-6. An Approach for Incremental Production and Local Personalization

Our Approach

The path we are going to take is straight forward. To fully understand VCP and thoroughly exploit the technology – do the fancy stuff, as Randy Pausch (2007) stated in his" Last Lecture" – we have to know the basics. We are going to look at concepts, mechanics, and demonstrations – basic tools, skills, and techniques, see how they fit together, and then see what we can do with them. We are not going to deal with fancy (and usually expensive) application software. The processes we are going to investigate are rudimentary, but very capable of producing a wide range of sophisticated results.

We will start by defining what a variable document may be. Then we will discuss various types of metadata before moving on to preparing the primary kinds of document content. Discussions on document design, workflow, tools and skills, and management processes will follow. We will conclude with a series of demonstrations on developing a number of variable documents using the knowledge we've been acquiring in the earlier chapters.

Suggested Reading

Bishop, B. (2008). *The big sort: the clustering of like-minded America is tearing us apart.* NY: Houghton Mifflin Company.

Godin, S. (2006). *Small is the new big.* NY: Penguin Group (USA), Inc.

Pausch, R. (2007). *The last lecture.* Retrieved September 6, 2008, from WXXI television, Rochester, NY.

3. What Is a Variable Document?

Discussion Points:

- ✓ Definition
- ✓ Standardization
- ✓ Variation
- ✓ Implications
- ✓ Further Reading

Definition

A static document is one in which every successive instance of the document is exactly the same. A variable document is a predefined design framework containing both static content and dynamic content. Dynamic (variable) content either varies over time between print runs or varies within a print run according to a stream of content that reflects a set of selection logic aimed at making each print impression (or set of impressions) unique for a given recipient, subject, or purpose.

There are many kinds of documents that can be termed "variable". That is, they can be produced with variable techniques. Some examples would be:

- Greeting cards and wedding invitations,
- Catalogues and marketing mailers,
- Business documents and contracts,
- Reports and academic papers,
- Poetry anthologies, cookbooks, and atlases,
- Newsletters, sales brochures, and restaurant menus.

Beyond content, purpose, look, and feel, these documents are different primarily because of when and how they vary. In terms of design concept, technology, and creative process they are similar. We can generally use the terms variable data printing (VDP) and variable content publishing (VCP) interchangeably, but both have their respective subtleties and implications – matters that don't concern us in our discussions.

What we are first concerned with is document structure and content organization, which we will refer to as "standardization". Secondly, we are concerned with when, how, and what in the document varies, which we will call "variation".

Standardization

When discussing variable documents, mentioning document standardization might sound like a contradiction. It is not. In fact, to facilitate the processing of variable content (thus making the document variable), the document framework must be highly standardized as also the structure of the content or content stream.

In talking about design standardization, we do not mean "relaxed standards", but rather very strict standards for the document. All components of the design are systematized so that they can ultimately come together automatically without human intervention. The intent is to build the first instance of the document in such a way that following iterations in a run, whether one or a thousand, can be generated entirely by the system (Figure 3-1).

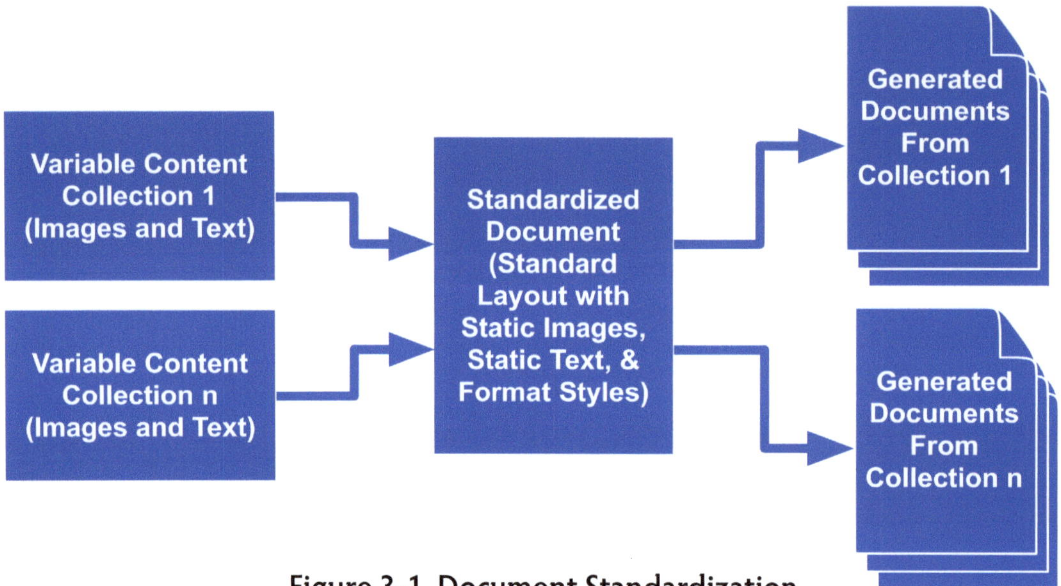

Figure 3-1. Document Standardization

After the document framework (including static content, document layout, and formatting styles) are established, then we can process one or more content streams coordinated according to document specifications, and produce completed document sets. We will devote a later chapter to discuss design setup details.

Note that standardization also affects the content stream. The content stream may represent a single version of a document, as in the case of a newsletter. It may also represent a set of repeating data elements used to generate a run of marketing mailers. In either example, desired results can only be achieved when the structure of the content stream and the structure of the document design correlate.

Variation

Variable documents tend to be associated with marketing and sales communications because the technology has been closely linked with customer relationship management (CRM). Each document generated in the print run is different in some respect in accordance with each targeted recipient.

Newsletters, on the other hand, are viewed as periodicals because they are produced on a regular frequency and all the documents in a run are the same. However, from a design and content management perspective, they are also variable documents – they just have different variation parameters, rhythms, and intervals. The process illustrated in Figure 3-2 could just as easily produce a run of personalized marketing mailers or publish a weekly newsletter.

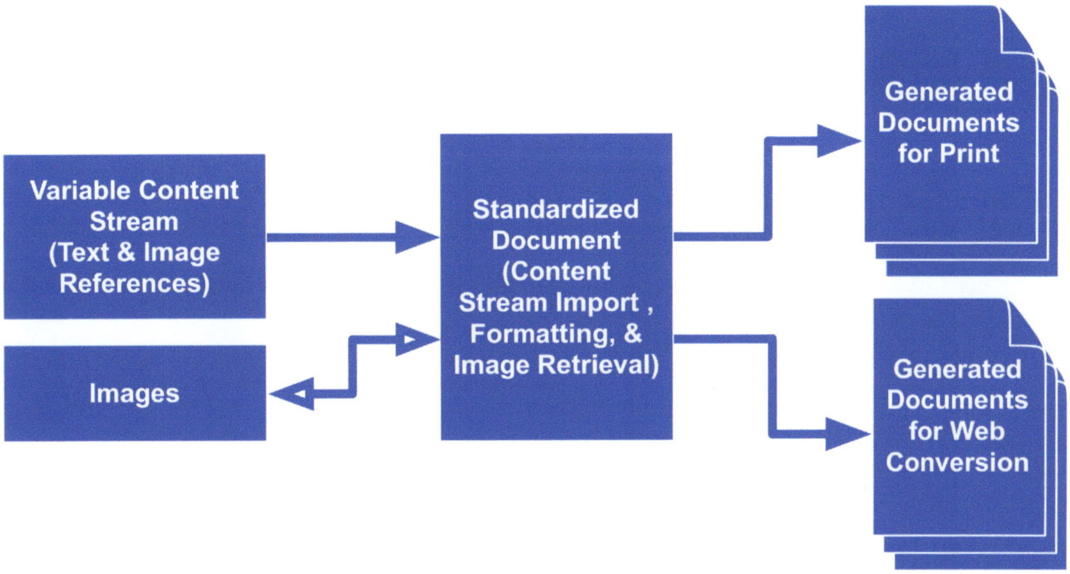

Figure 3-2. Document Variation

In considering the document as a whole, what is the frequency of the variance? Does variation take place within a single production run or between runs? If between production runs, what is the variation interval – daily, weekly, monthly, or some other frequency?

After determining frequency, identify the individual design elements within the document that will vary. For each variable element, describe the cause and the result of each change. Here are the primary considerations (there may be others):

1. What content elements vary?
2. When does each content element vary?
3. Why does each content element vary – what is the purpose behind the variance?
4. How does each content element vary? Does it change color, shape, wording, or is it completely replaced?
5. What decision criteria and logic is applied to determine and initiate each variance?

It is worth noting here that many VCP applications define the variations as part of the document design process. The variation logic resides and is executed within the design. That is not the approach taken here.

Rather, we take a definite data-centric approach that sets up any variances as part of content preparation; after all, we are talking about content-driven printing and publishing. This approach will be covered in Chapter 7, "Desktop DBMS – Managing the Source Data" , and Chapter 8, "Techniques – Managing the XML Content Stream".

Implications

There are a couple of ramifications in this chapter for tools and skills that we need to keep in mind. Creating a variable document is not strictly an artistic process. It is also an engineering process. This means the artisan may be faced with some constraints to artistic license, and the engineer is going to need some artistic awareness.

Art and engineering have their contradictions. The artist-engineer or engineer-artist is the best recipe for success, depending on which (art or engineering) is thought most critical in producing desired results. There's little to be gained in arguing; who knows for sure what the perfect mix looks like? Maybe in a perfect world, but we all know how realistic that is.

So, we need to compensate for potential imperfection. (Easier said than done!) There are four possibilities we might take either individually or in some combination:

1. Simplify the development tools,
2. Standardize the development environment and work process,
3. Engender flexibility within the standard workflow,
4. Simplify the end product – something to be avoided if at all possible.

Further Reading

Maivald, J. & Palmer, C. (2008). *A designer's guide to Adobe InDesign and XML.* Berkeley, CA: Adobe Press, Peachpit.

Powell, G. (2007). *Instant InDesign: designing templates for fast and efficient page layout.* Berkeley, CA: Adobe Press.

TWO: Foundation

There are three types of digital content that may come into play with variable content documents: graphics, data, and text. The aim here is to assemble a foundation of knowledge for creating variable documents. We don't need to know everything about each subject to be successful – only a limited subset. That's all we'll need to understand and ultimately manipulate content to create variable results.

4. Metadata About Digital Assets

We begin our discussion on document content by focusing on digital assets – primarily graphics and imagery. The purpose of the chapter is to discuss why a "picture" without context does not provide an adequate understanding to effectively use it. Content requires definition, description, classification, and organization to improve a designer's ability to access, select, maintain, and use available graphics.

a. Overview
b. File Data
c. Camera Data
d. IPTC Core Data
e. Other Data Categories
f. Search, Identification, & Selection
g. Some Explanatory Reading

5. Metadata About Data

This chapter continues in the flavor of the prior discussion, but focusing this time on data. The chapter describes the importance, actually the necessity, of understanding the underlying structure, organization, and purported quality of data collections. That knowledge directly facilitates accessability, insight into processing requirements, appropriate and successful use.

a. Overview
b. Data Files
c. Databases
d. Recommended Reading

6. XML – Metadata About Text

Chapter 6 concludes the metadata discussion by addressing text from a structural perspective. Although XML is normally not referred to as metadata, never-the-less, it does provide a descriptive method and tool for understanding and describing the structure (organization) of text. The chapter surveys those features of XML that a variable document developer is liable to need.

a. Overview
b. Building Blocks
c. Well Formed
d. Valid
e. Keep In Mind...
f. Recommended Reading
g. Supplemental Information

4. Metadata About Digital Assets

Discussion Points:

- ✓ Overview
- ✓ File Data
- ✓ Camera Data
- ✓ IPTC Core Data
- ✓ Other Data Categories
- ✓ Search, Identification, & Selection
- ✓ Some Explanatory Reading

Overview

What is metadata? Simply put, it is data about data – in this case digital images. When we talk about metadata in later chapters on data and text files, or the text itself, the definition still holds true.

Images and other file types carry associated explanatory information. Even a simple text file will carry details such as file type, creation and modification dates, and file access permissions. The photograph from a digital camera will provide that information plus camera settings, color profile, and much more.

For this discussion, metadata can be categorized as technical, descriptive, and contextual. A person who wants to use an image first needs to find it and then determine how best to use it (if at all). In other words, the user of a given image file needs to understand who, what, when, where, how, and why about the physical file and the content of the file – the image.

Most metadata is organized into logical groups. Our examples are based on how Adobe Bridge categorizes and presents metadata for use within their graphic design toolkit. Adobe Bridge is a good introductory point for DAM (Digital Asset Management), and the knowledge acquired in using it is transferable to larger systems. Figure 4-1 shows a digital image (the "farmhouse" photograph) as the basis for the examples in the remaining chapter discussion.

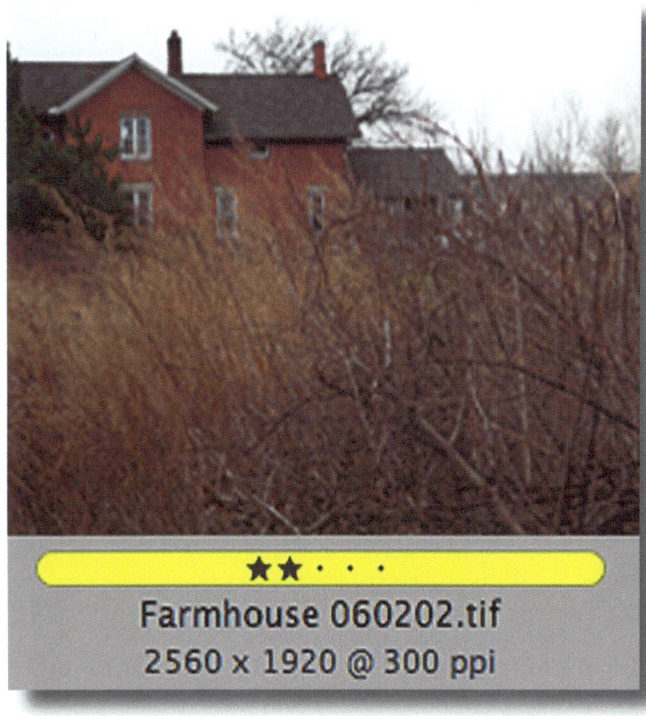

Figure 4-1. A Photograph in Adobe Bridge

File Data

File metadata pertains to the content container as opposed to the content itself. It tells what type file the container is, what application originated the file, its size (how much space it occupies), and tracking details (creation and modification dates). Files containing image content include image resolution, image dimensions, and color profile information. Figure 4-2 shows file metadata for the farmhouse photograph as displayed by Adobe Bridge.

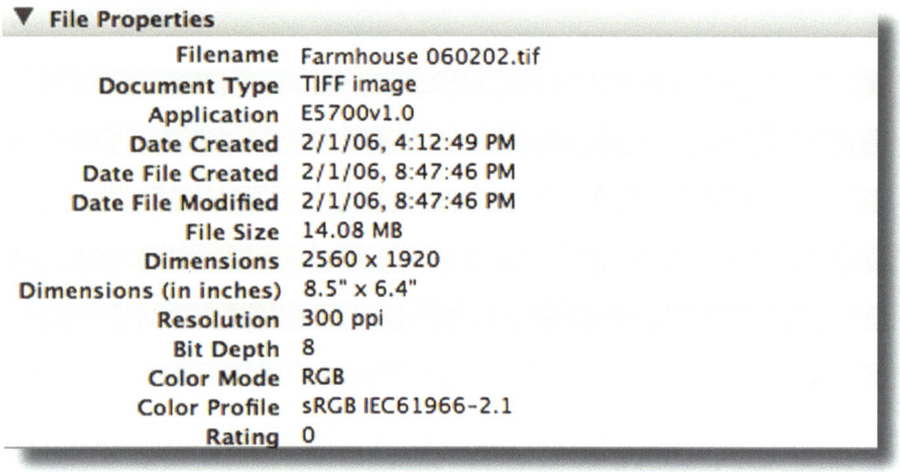

Figure 4-2. Adobe Bridge File Metadata

Camera Data

Camera metadata provides information about image acquisition (the capture device, its settings, and processes). This information is important in understanding the image from a technical perspective, and provides the basis for correction, enhancement, or modification of the image. Figure 4-3 displays camera metadata for the farmhouse photograph.

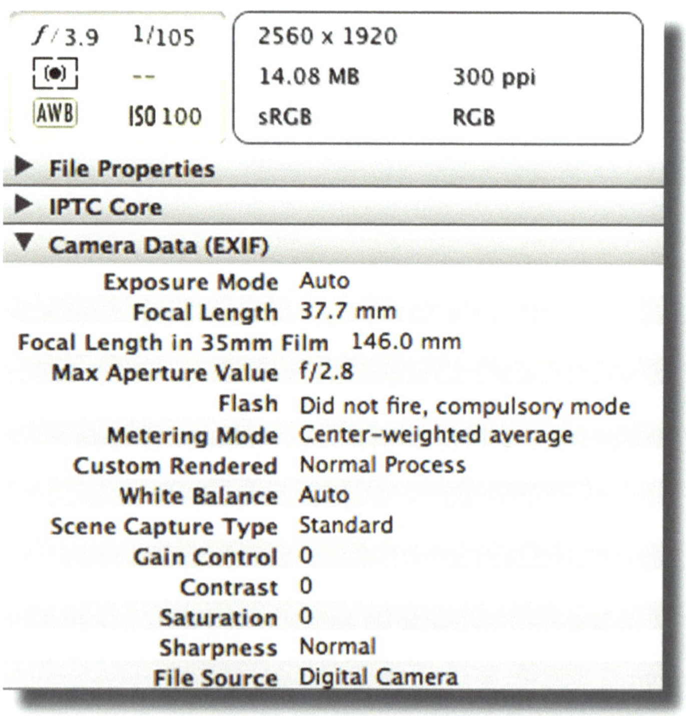

Figure 4-3. Adobe Bridge Camera Metadata

IPTC Core Data

IPTC stands for International Press Telecommunications Council. This is journalistic core data about the image (who, what, when, and where).

As displayed in Figure 4-4, it furnishes location and contact information regarding the image creator, describes the image, explains where it was acquired, and also provides project and copyright information.

Notice that there is allowance for a considerable number of categories and details. Only a few have been used. Since use of these items is primarily optional, the antidote for non-use is for the organization to establish an image standard that details the minimal set of metadata that will be provided with each image.

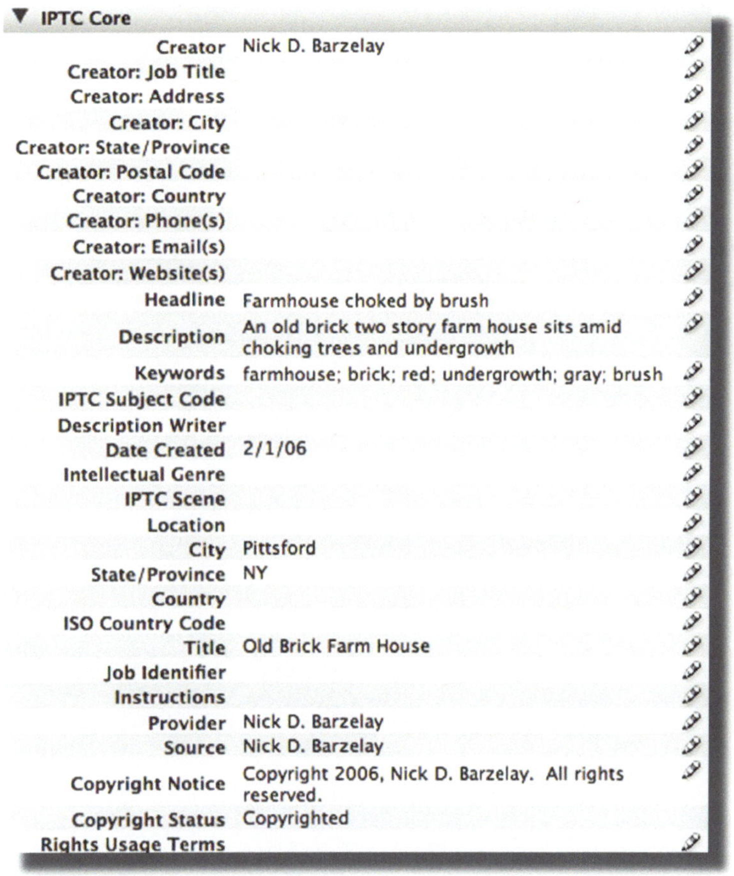

▼ IPTC Core

Creator	Nick D. Barzelay
Creator: Job Title	
Creator: Address	
Creator: City	
Creator: State/Province	
Creator: Postal Code	
Creator: Country	
Creator: Phone(s)	
Creator: Email(s)	
Creator: Website(s)	
Headline	Farmhouse choked by brush
Description	An old brick two story farm house sits amid choking trees and undergrowth
Keywords	farmhouse; brick; red; undergrowth; gray; brush
IPTC Subject Code	
Description Writer	
Date Created	2/1/06
Intellectual Genre	
IPTC Scene	
Location	
City	Pittsford
State/Province	NY
Country	
ISO Country Code	
Title	Old Brick Farm House
Job Identifier	
Instructions	
Provider	Nick D. Barzelay
Source	Nick D. Barzelay
Copyright Notice	Copyright 2006, Nick D. Barzelay. All rights reserved.
Copyright Status	Copyrighted
Rights Usage Terms	

Figure 4-4. Adobe Bridge IPTC Core Metadata

Other Data Categories

As expected, metadata is also recorded for video, audio, and medical imaging. The types of data provided are all dependent on the acquisition technologies used and their respective capabilities. For instance, satellite images may contain global imaging system (GPS) information. DICOM (Digital Imaging & Communications in Medicine) would describe medical images.

The important point is that whatever digital asset management system is employed or whatever approach is used, it needs to comprehend the existence of a variety of metadata and make it accessible to its content users.

Search, Identification, & Selection

Before using a file, a person has to find it. Before being able to select just the right image to meet project criteria, a person first needs to locate a group of files that in some sense satisfies that criteria. One common solution is to use descriptive file names. This may work for small file collections, but it has limitations that may produce unsatisfactory long term results, such as escalating file management problems (including file mix-ups, misplaced files, and accidental file destruction).

File Naming Conventions

One remedy for the shortcomings of descriptive file naming is to use file naming conventions that eliminate the randomness of descriptions. Naming conventions should provide individual file identification (uniqueness) and the capability to identify related files (group selectivity). Additional information built into a file name may be ownership, version, and date. We are talking about long names in a fixed sequential pattern with a fixed format and standard terms and acronyms.

While using naming conventions is an important step, it does not provide all the information or functionality needed. In fact by themselves, they may ultimately cause confusion and dysfunction. Implementing standards usually means implementing assisting technologies – in this case, digital asset management. File naming will likely need additional search techniques.

Key Words – A Controlled Vocabulary

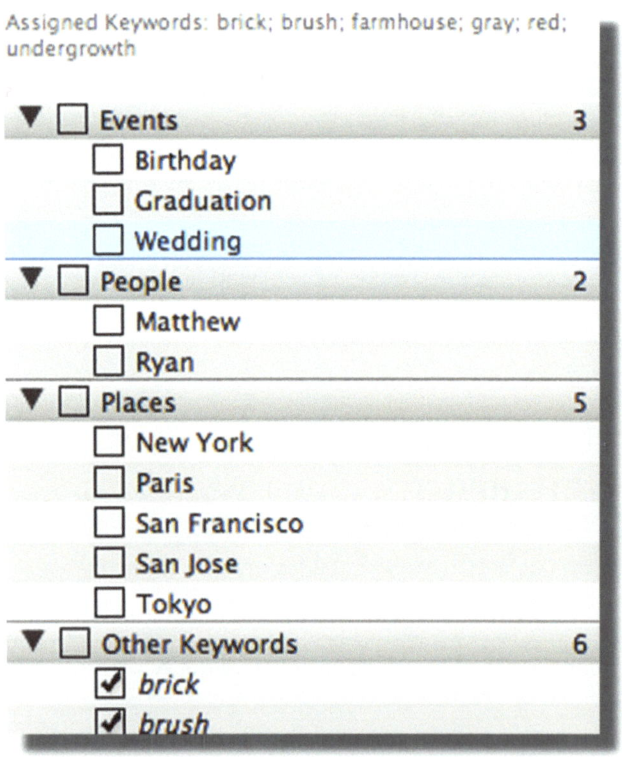

Figure 4-5. Key Word Grouping in Adobe Bridge

As a metadata component, keywords assist in locating and selecting desired images by standardizing the set of available descriptive word choices. Adobe Bridge provides an easy way to build and manage an outline of key words (Figure 4-5).

The controlled vocabulary is one more aspect of implementing standards. The idea – and the challenge – is to build an outline (taxonomy) of features and attributes facilitate grouping digital assets with common characteristics, yet identify ingindividual assets according to some unique set of qualities. Such a scheme may be the basis for setting up digital asset storage.

However, some experts advocate "freely descriptive" key words, but arbitrary word choices invest a lot of trust that one person is likely to use the same terminology as another. More dubious is the assumption that the next person wanting to find a particular image will be able to guess the first person's word selections. The passage of time, changes in personnel, and an increasing number of digital assets seriously limits the likelihood this approach will continue to produce desired results.

Labeling & Rating

To assist in classifying and searching for digital assets, some DAM systems provide built-in supplementary capabilities. Adobe Bridge is no exception. It provides two tools that may be used separately or in tandem. One is a color labeling scheme; the other is a grading scheme. Figure 4-1 a few pages back exhibits a color label and rating below the image.

Color labeling provides six alternatives: Red, Yellow, Green, Blue, Purple, and None. By themselves, the colors are meaningless. It is up to the user to determine what each color will mean. Rating (or grading) is another matter. It is based on a "star" system ranging from none to five stars: "Not Rated", "Rejected", and "★" up to "★★★★★".

Again, it is up to the user to determine what is being rated and the granularity of the rating. Rating could be based on image usability, quality, popularity (how many times it has been used), or something else. That something else could be a more sophisticated classification scheme with the color label signifying a category, and the stars indicating rank within category.

Filtering

Once metadata, key words, labels, and ratings have been applied, there needs to be a way to put them easily to use. One way is by submitting search arguments to the DAM or file system. For example, the argument might be, "select all images in URI (Uniform Resource Identifier) where keyword = 'house' and label = 'green' and rate = 'four' and ISO >= 100". This is very much like a database query. Each system will have its own search language and syntax. There has to be an easier way!

The advantage of using a DAMS is that it will probably have a built in mechanism or tool for simplifying search and selection. Adobe Bridge provides a filter (Figure 4-6).

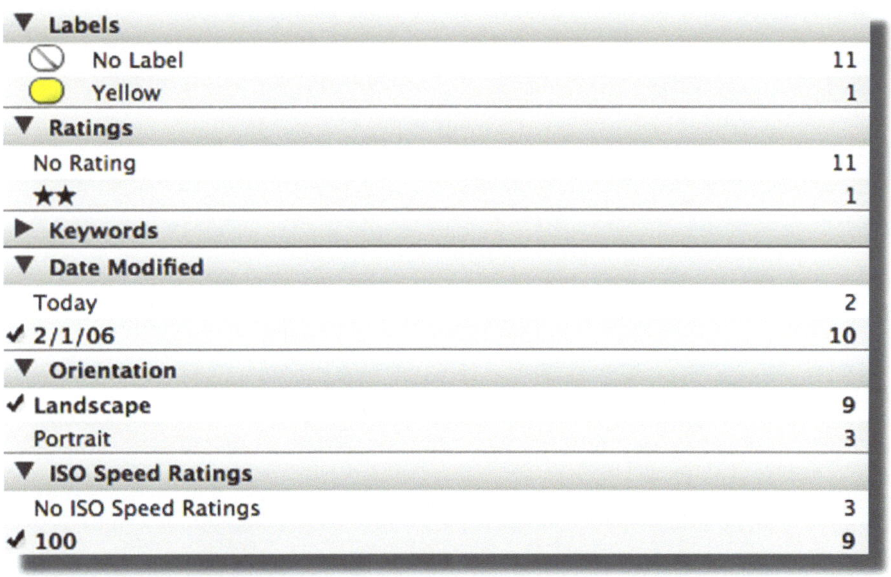

Figure 4-6. Adobe Bridge File Filtering

The filter is a display-based tool that uses a common set of selection categories. It "filters out" everything that does not meet the preset criteria and provides an accounting of all the images that match the selection criteria. The user can select a category or subcategory and the filter will show any matching images that meet selection criteria.

Some Explanatory Reading

Information presented in this chapter comes from a hands-on exploration of Adobe Bridge, the desktop digital asset management module packaged with Adobe Creative Suite. First-hand knowledge of this software can enhance understanding this chapter and subsequent chapters that allude to DAM concepts.

Other information came from the manual and from experience with a digital camera, an older (2002 vintage high-end amateur) Nikon. Most, if not all, digital cameras collect information about each photo, which subsequently downloads along with the digital image. This is worth closer examination, even if you only use a digital camera for personal photos.

There are other ways to search and find files with desired content beside digital asset management. For those who are overly technically inclined, not in a hurry, or those who enjoy a challenge, there are operating system level search utilities that can be run from the terminal of either a PC or Mac. The reference below is just one of many sources for the person who really gets enjoyment from struggling and frustration, or perhaps from just plain mental self-abuse.

Taylor, Dave. (2005). *Learning Unix for Mac OS X Tiger.* Sebastopol, CA: O'Reilly Media, Inc. (or similar topic).

5. Metadata About Data

Discussion Points:

- ✓ **Overview**
- ✓ **Data Files**
- ✓ **Databases**
- ✓ **Recommended Reading**

Overview

Data is "the" critical component of VCP design and development. The reason metadata is important is that it significantly assists the developer in thoroughly understanding data used in the development workflow.

The origination points for most data may be a database, a collection of application files, spreadsheets, or a compilation of data stored sequentially on disk or some other media (maybe, magnetic tape). What the VCP developer usually wants is not the source data, but a prepared extract from the source data. Ideally, the database extract is performed by the data provider, the sanctity and security of the database is maintained, and the VCP developer is spared the complexities and peculiarities of the source data collection.

This extract will probably arrive as some sort of data file, spreadsheet, or possibly a single table desktop database. This extract may be of dubious quality and acuracy. The VCP developer has to put the submitted data into an easily workable form and ultimately integrate it into the document design.

We are going to discuss the concepts, structures, and elements involved in this process from the perspective of metadata. This information will be useful for Chapter 6 where we introduce XML (extensible markup language), and Chapters 7 and 8 where we discuss the processes of content stream preparation.

Data Files

What is a data file? It is a file that is read sequentially from start to finish. It has a series of repeating data elements organized in some manner into groups (records containing a set of fields). Figure 5-1 (next page) illustrates this arrangement. The illustration is a nicely organized sequence of data that would seem easy enough to process. In reality, processing may not be so easy.

The physical layout of the data may be nothing more than a stream of characters (worst case). Then again, data could be grouped into logical records using an open space (called an IRG – inter record gap). Data may be grouped into records using nothing but a semicolon (";"), carriage return symbol (CR), line feed symbol (LF), or a combination (CRLF). These alternatives will be much easier to manipulate than the unarticulated character string.

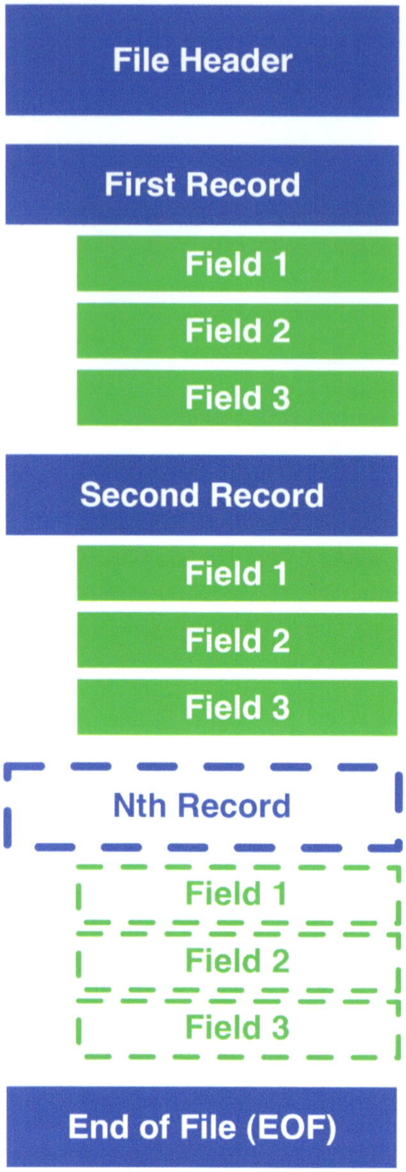

Figure 5-1. A Sequential Data File

Within each record, data fields may be strung together (a character string) or separated by some designated character – traditionally called a CSV (comma-separated values) or delimited file. The separators may be a designated punctuation mark rather than a comma.

Using punctuation marks in general and commas in particular to separate field values does have a downside; they may reasonably occur within the data or have conflicting usages (a semicolon as both field and record separator). The best selection for separating fields is the tab space (tab-separated values).

However, the primary problem with data files is that they include within the file itself very little metadata. With the exception of the header record which describes the file, there is nothing embedded in the file that describes the structure of its content. This structural information is noticeably absent in the visual presentation of a text file's header (Figure 5-2 below).

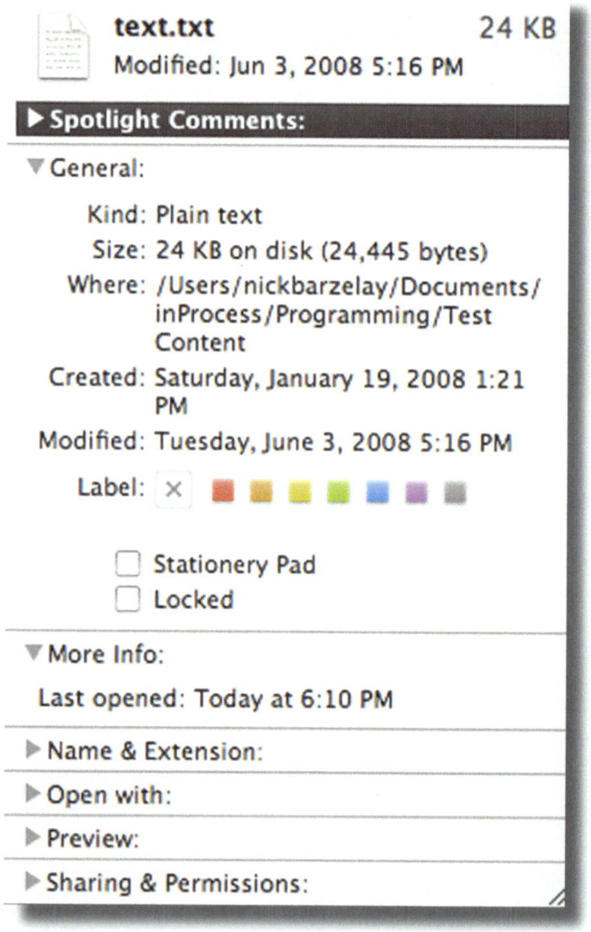

Figure 5-2. Mac OS X File Header Panel

Unless someone ships a detailed file layout document with the file, the only resort is to examine the file. This may not be a trivial exercise, because the quality and effectiveness of VCP documents is directly dependent on the quality of the source data. Data files that are not delimited will probably require programmatic preparation, meaning the expenditure of additional time and effort.

Files such as CSV (comma separated values) or tab-delimited files are also processed sequentially. The advantage is that fields within each record are easier to differentiate and can be of varying character length. They essentially form a matrix of data (shown in Figure 5-3) that can include column names as the first row. Their disadvantage is that they may still require programmatic support if processed "as is".

However, delimited files can easily be imported into a spreadsheet (such as Microsoft's Excel) or directly into a desktop database. Conceptually, Figure 5-3 with the delimiters removed could easily be a model for a spreadsheet or database table (see Figure 5-6 further below).

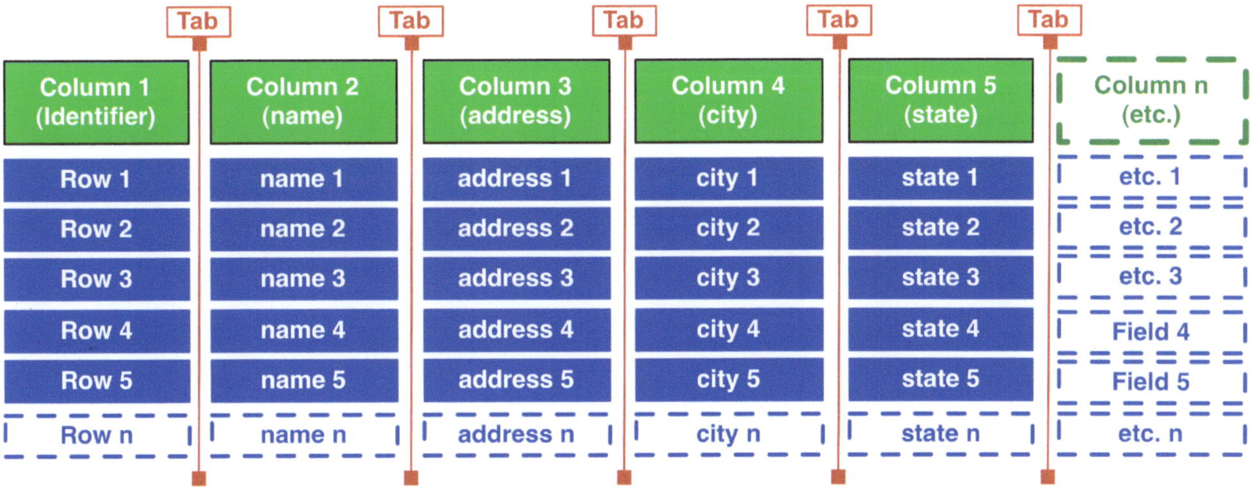

Figure 5-3. A Tab-Delimited Sequential Data File

At the very least, the spreadsheet will provide structure to the data and a way to examine and manipulate it visually (something the delimited file can't provide). However, some programatic preparation within the spreadsheet may still be necessary. Data saved as a spreadsheet is usually easy to import into a desktop database for XML conversion, or exported back out as a tab-delimited file for direct conversion into an XML data stream (discussed in a latter chapter).

This potential merry-go-round of file movement between formats may sound crazy, but it is easier to move files between friendly environments than work with them "as is" in a hostile format.

Databases

Given the dearth of metadata in sequential files along with associated data manipulation issues, importing the data into a database management environment to prepare for content stream generation and document integration is an important step. Why is that? Because the database management environment provides the metadata and tools to easily understand and manipulate both data structure and data content. After all, the entire project rests with the data. Even if specialized VCP application software is in use, data preparation is necessary.

Let's leave data processing to Chapters 7 and 8, and continue discussing metadata. Just as in sequential files, there are two categories – one to describe structure and the other to describe content. The chief advantage with databases is that they contain both types. Both categories are physical manifestations of the database schema. The schema is the blueprint of the database. It can be generated directly from the database, rather than being a separate document.

Structure

Database structure is essentially one or more tables made of rows and columns. A single table database is often referred to as being "flat". It is somewhat like a spreadsheet but oriented toward manipulating data as opposed to mathematical modeling. In multi-table databases, the individual tables do not have to connect together, but they usually do. That connection is called a "relationship" – hence relational databases.

There are three standard relationships. The first is a peer-to-peer relationship, which is infrequently used because, from a design and functionality perspective, why add the overhead of processing two tables when one table would do. The second is a one-to-many relationship. A parent table has one or more child tables, and for each unique instance in the parent table, there can be multiple instances in the child table which point back (are related to) that single parent instance. Conceptually, it looks like Figure 5-4. Note that a child table may be the parent of other child tables; a parent table may be the child of another table up the hierarchical ladder.

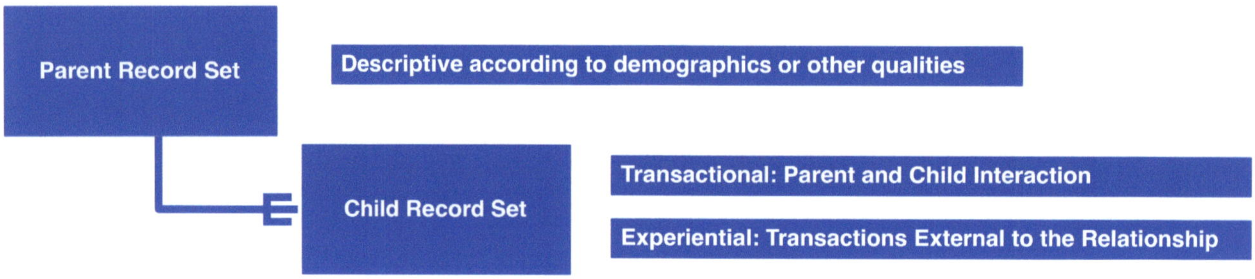

Figure 5-4. Relational Data Model

This last point is shown in Figure 5-5 (directly below), which illustrates a physical rendering of the concept – in this case, a set of one-to-many relationships in a FileMaker database. A customer goes shopping at different stores (activities), and at each store buys a number of items (details).

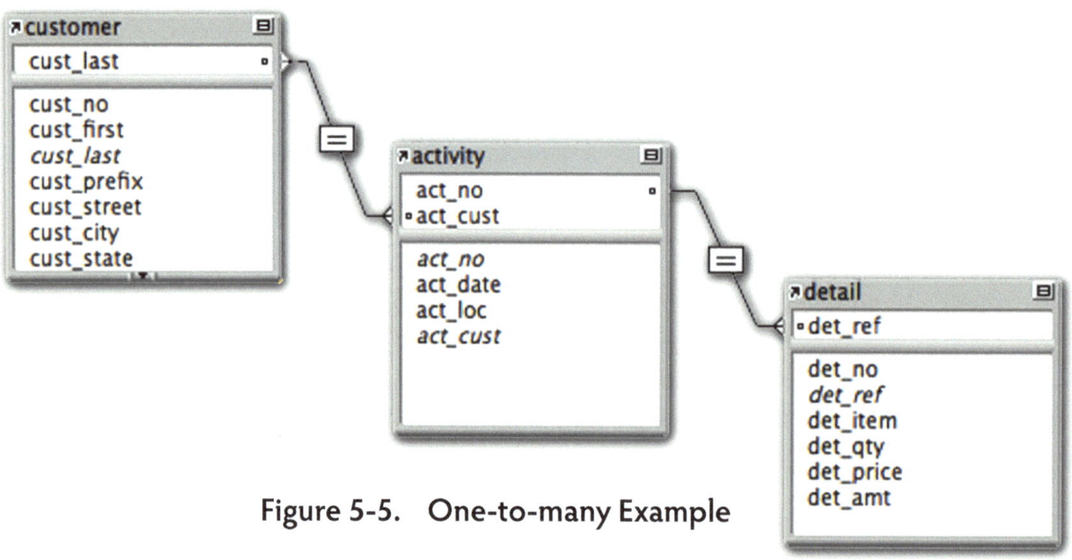

Figure 5-5. One-to-many Example

The third is a many-to-many relationship, for example, multiple people own multiple makes of cars. This type of relationship is more complex than what it might initially seem. You might think there are two connected tables: one filled with people who own cars and the other filled with the different makes of cars.

Easily overlooked is the need for a third table, called a "junction table" that describes ownership between car makes and individuals. What this represents is actually two tandem one-to-many relationships: an individual may own multiple cars, and a particular make of car is probably owned by multiple people.

Databases are files, so they contain metadata commonly associated with file headers. In addition, they contain metadata related to the database structure:

- Table name
- Table relationships (parent and child table names)
- Relationship types and their direction (the "one" or "many")
- Table unique key (making every record in a table unique)
- Table foreign key (points back to the parent from the child)
- Number of fields (columns)
- Number of records (rows)
- Table access (accounts and privileges)

Content

Tables are the content containers of databases. Tables contain rows and columns. The rows represent individual records and the columns represent fields within the rows (similar to the records and fields of a sequential file except treated as a matrix). Figure 5-6 illustrates this idea conceptually.

Column 1 (Identifier)	Column 2 (name)	Column 3 (address)	Column 4 (city)	Column 5 (state)	Column n (etc.)
Row 1	name 1	address 1	city 1	state 1	etc. 1
Row 2	name 2	address 2	city 2	state 2	etc. 2
Row 3	name 3	address 3	city 3	state 3	etc. 3
Row 4	name 4	address 4	city 4	state 4	Field 4
Row 5	name 5	address 5	city 5	state 5	Field 5
Row n	name n	address n	city n	state n	etc. n

Figure 5-6. Database Table Model

Adhering to the relational constructs presented previously, Figure 5-7 on the following page shows the field (or column) specifications and pertinent metadata for a parent table as presented in FileMaker Pro. Figure 5-8 below shows similar specifications and metadata for the related child table.

Field Name	Type	Options / Comments (Click to toggle)
cust_no	Number	Auto-enter Serial, Can't Modify Auto
cust_first	Text	
cust_last	Text	
cust_prefix	Text	
cust_street	Text	
cust_city	Text	
cust_state	Text	
cust_zip	Text	
cust_val	Number	
cust_type	Text	
cust_grp	Text	

Figure 5-7. Parent Table Example

Field Name	Type	Options / Comments (Click to toggle)
det_no	Number	Auto-enter Serial, Can't Modify Auto
det_ref	Number	
det_item	Text	
det_qty	Number	
det_price	Number	
det_amt	Number	

Figure 5-8. Child Table Example

Content-related metadata describes and prescribes the types of data contained in a given table. Different DBMS's approach data definition and typing in varying ways, although conceptually there is considerable similarity. Desktop databases simplify data typing, and integrate display and reporting capabilities with data handling and programming. Their metadata may diverge in some instances from what is found in larger systems. However, they provide a satisfactory appreciation for basic details. Below is a list of common metadata elements found in most desktop and larger scale databases:

- Column name
- Data type (text, number, date, time, or binary container) and Size (number of characters)
- Use (primary key, foreign key, index key, or content container)
- Creation date and Modification date
- Permissions (who can read or manipulate field content)
- Validation value (a specified test value)
- Default value (including automatic incrementing and increment value)
- Data entry mask and display format (formatting assistance for data input and retrieval)

Recommended Reading

The three texts below are dated – more recent editions are available – but they will still provide the reader an appreciation for basic database design. Microsoft Access is a PC-only application. FileMaker is cross-platform compatible (Mac and PC). Either is a good choice. In addition there are a number of extremely good SQL (Structured Query Language) databases with varying features and capacities..

Coffey, G. & Prosser, S. (2005). *FileMaker Pro: the missing manual.* Sebastopol, CA: O'Reilly Media, Inc.

Kofler, M. (2005). *The definitive guide to MySQL 5, third edition.* Berkeley, CA: Apress.

Roman, S. (1997). *Access database design & programming.* Sebastopol, CA: O'Reilly & Associates, Inc.

6. XML – Metadata About Text

Discussion Points:

- ✓ Overview
- ✓ Building Blocks
- ✓ Well Formed
- ✓ Valid
- ✓ Keep In Mind...
- ✓ Recommended Reading
- ✓ Supplemental Information

Overview

Most non-technical commentary on XML (extensible markup language) treats it as another tagging language. It has a number of forerunners, most notably SGML (standard generalized markup language) and HTML (hypertext markup language) and its derivatives – DHTML and XHTML (dynamic and extensible versions of HTML). Unlike XML. these tagging languages define how a document "looks". XML describes the structure of a document's content.

The other markup languages generally have a fixed set of tags to accomplish specific formatting requirements and design functions associated with particular document display applications (for instance, different Web Browsers). XML doesn't have a fixed set of tags, since it doesn't have to accommodate any format processing requirements. It only has to describe the structure of particular content.

The person responsible for XML tagging is free to either use pre-packaged sets of descriptive tags defined for a specific purpose (for instance, "DocBook" or "MathXML"), or create a customized tag set pertinent to a single document (like a formal paper), or to a class of documents (like newsletters or marketing collateral). Hence it is extensible text-related nomenclature that can be specialized to meet differing document definition requirements.

Since XML defines document textual content – its structural characteristics and associated attributes – we've termed it "metadata". Because it only describes content structure, XML-tagged text can be styled with separate non-XML formatting instructions to satisfy multiple purposes. This capability makes XML an excellent mechanism for integrating the same text into a variety of variable documents (for instance, breakfast, lunch, and dinner menus, a take-out menu, a menu for a Web site, and restaurant advertisements).

XML has a considerable array of descriptive and structural processing facilities, a factor that makes the language appear intimidating and overly complex. However, from the perspective of creating variable content documents, there is good news. We don't need to know all the "in's and out's" of XML to be productive. In fact, it's better to simplify the XML content stream (eliminating extraneous XML structural elements and unnecessary white space inserted for readability. Notice that size, quantity, and substance are not affected by simplification. Our learning approach will cover three areas: (1) the basic XML we'll be using later, (2)a small completed XML example, and (3), some additional critical details.

Building Blocks

The constructs discussed in this section are ones relevant to constructing and using content streams tagged with XML. Consequently, this is not a full tour of XML, nor is it necessarily a full explanation of each expression described below. It only explains essentials related to their use in our particular application. The examples and code listings are patterns that you can modify and adapt. *(If you have other applications where you might want to use XML, you'll need to supplement this treatment with additional reading. A good start is the suggested material at the end of this chapter.)*

XML Declaration

The very first line of any XML document must be the XML declaration. It identifies the document type as XML. For our purposes, version 1.0 is a safe bet. It is in its fifth edition, indicating that issues are being addressed. Version 1.0 is limited to unicode version, UTF-8 (Unicode Transformation Format). If more robust unicode is needed, the XML version will be 1.1.

```
<?xml version="1.0" encoding="UTF-8"?>
```

Namespace Declaration

The namespace declaration qualifies (or puts in context) an XML element. The statement below is for accessing an XML schema. Though we will not be using schemas, you may encounter this declaration, and may need it at a later date or for another application.

```
<xs:schema xmlns:xs="http://www.w3.org/2001/XMLSchema">
```

However, you may encounter other uses for namespace – for instance with XSL (extensible stylesheet language) used as processing instructions for XSL transformations (XSLT) discussed later.

```
<xsl:stylesheetxmlns:xsl="http://www.w3.org/1999/XSL/Transform" version="1.0">
```

Processing Instruction

The instruction tells an XML application how to process an XML document using a non-XML file. The case below calls for processing a document type definition (DTD) in "filename.dtd" starting at the document root, "rootname". Note that the DTD isn't really an XML construct.

```
<!DOCTYPE rootname SYSTEM "filename.dtd">
```

The next example calls for processing a cascading style sheet (CSS – which also isn't an XML document) of type "text/css" located at pathname (or URI – uniform resource identifier) in a file called "filename.css".

```
<?xml-stylesheet type="text/css" href="path/filename.css">
```

Comment

The comment is usually descriptive or instructional. For variable content publishing, keep comments out of the data or content stream. While this is not an XML caveat, it is prudent to keep the stream, as uncluttered as possible. Note the example comment below:

```
<!-- the double dashes are not single hyphens -->
```

As warning, do not substitute hyphens for the double dashes.

Element

Element is the basic XML building block. It is composed of a start tag, content, and an end tag.

```
<any_name>some element content</any_name>
```

Most content definition, preparation, and processing will almost entirely involve elements (particularly for VCP). Elements may contain elements. In such cases, they may be indented for readability. The example below is an element containing imbedded elements (sub elements).

```
<state>
    <name>New York</name>
    <abbreviation>NY</abbreviation>
</state>
```

Indenting is not needed for processing. Indenting creates white space which may require extra work when populating a document design with the XML. The best approach for processing is to keep all elements left-aligned.

```
<state>
<name>New York</name>
<abbreviation>NY</abbreviation>
</state>
```

Elements may be empty – contain no content. Empty elements can be shown in two ways: simply start and end tags with nothing between (first example), or as a combination start and end tag like the second example (which is preferable).

```
<tag_name></tag_name>
<tag_name/>
```

Attribute

An attribute is part of the element tag, not the element content. It qualifies the element or provides further element-level information.

```
<element attribute="item">element content</element>
<state abbreviation="NY"> New York</state>
```

An attribute can also reference (href) an external file to provide added content without impacting the size of the XML content stream. This is important if some content is stored at a remote location.

```
<image href="file://server3/photographs/RedRose.jpg"/>
```

As used in the content stream, attributes can simplify the XML by keeping content at the element level, and reserving attributes for pulling external files such as digital images into the design (as demonstrated by the highlighted XML below).

```
<state>
    <name>New York</name>
    <abbreviation>NY</abbreviation>
    <map href="file://NYmap.jpg" />
    <population/>
</state>
```

Predefined Entity

Some common symbols are part of XML syntax. If any exist in the content, they need to be redefined. XML provides a mechanism for doing this – the predefined entities shown below. The predefined entity can be applied with a "search and replace" from a simple text editor.

&	ampersand (&)
<	less-than or left angle bracket (<)
>	greater-than or right angle bracket (>)
"	quote marks (")
'	apostrophe or single quote marks (')

General entities (not predefined) can be defined by the developer to substitute a value into the content where the value in and of itself might not correctly parse (assessed, interpreted, or comprehended) when encountered by the XML processor. Like the predefined entities above, the general entity can be applied with a "search and replace" from a simple text editor. The general entity is declared in the DTD in the pattern, *<!ENTITY entity_name "substituted value">*. Here is an example showing the declaration, the XML statement, and the parsing result (highlighted):

```
<!ENTITY bang "!">
<warning>Hazardous Material&bang; Do Not Ingest&bang;</warning>
```

Hazardous Material! Do Not Ingest!

XML Example

Listing 6-1 is an example of a small XML document – a lunch menu for the work-week. Ok, part of a week! Anyway, this document is an easy-to-understand portrayal of how XML is constructed and what it might look like.

```
<?xml version="1.0" encoding="UTF-8"?>
<!DOCTYPE lunch SYSTEM "SackLunch.dtd">
<lunch>
        <day name="Monday">
            <sandwich>Ham &amp; Swiss on Rye</sandwich>
            <fruit>Orange</fruit>
            <chips>Potato Ruffles</chips>
            <drink>Root Beer</drink>
        </day>
        <day name="Wednesday">
            <sandwich>Egg &amp; Olive on Wheat</sandwich>
            <fruit>Apple</fruit>
            <chips>Corn Chips</chips>
            <drink>Iced Tea</drink>
        </day>
</lunch>
```

Listing 6-1. Short XML Example (<lunch> is the root)

Well Formed

There aren't a lot of rules and syntax to learn in XML. Different authors provide slightly different lists. Essentially, there are only nine rules to worry about. That's not a lot, especially when compared with other technologies.

If any XML document adheres to these nine conventions, it is considered "well formed". That means the document does not contain any syntax errors. Here's the list:

1. XML documents must begin with an XML declaration
2. An XML document must contain a single root element that contains all the other elements
3. All elements must be nested and not over-lap (child tags are closed prior to closing parents)
4. XML tags are case-sensitive and need consistency
5. Tag names can't contain spaces or start with "XML", a number, or punctuation (except the underscore)
6. All elements require start and end tags (empty tags may contain the end marker)
7. All attribute values must be enclosed in quote marks
8. Syntactical characters are replaced with pre-defined entities
9. The conventions apply to all types of XML documents

Valid

What is a valid document? A valid document is one that is first "well formed", and second, in complete agreement with the document specification. The specification can be expressed as an "XML Schema" (a type of XML document), or it can be defined with a document type definition (DTD). We have no need to use the schema in this application of XML, although that could easily change at a later time.

The DTD is a non-XML document in that it is not constructed with XML statements. It is built with its own syntax and vocabulary. It describes a document's structural hierarchy as a descending cascade (demonstrated later). During validation, the document processor compares the XML document to the DTD and either approves the XML or displays errors.

Why does a document file need validation? Isn't being well formed enough? No! The reason is that standardization provides an advisable level of quality assurance. Validating insures that all XML content structures used in a particular application meet specifications. Suppose XML representing an extract from 2,000 customer records will be used to generate a run of personalized variable documents. Validation will determine if all 2,000 records are structurally the same. A defect in one record (number 400 for example) may affect placement of the remaining 1,600 records. Automated document generation based on standardized content and reusable design templates needs preprocess validation to insure accurate design/content integration.

Consider this question! Do you want to "eyeball" 2,000 records? Suppose each record contains ten elements. That's 20,000 item you'll need to examine. Could you miss something or overlook a problem? Possibly, maybe probably! I'm not a big fan of self-induced extra work myself.

DTD/XML Connection

The DTD can be paired with the XML document in one of two ways. The document definition can be embedded in the XML document file after the XML declaration as an "internal DTD". The complete document definition is contained within brackets, as highlighted below. The problem with this approach is that it defeats the idea of separating functionality by purpose, which complicates or prevents XML and DTD reuse in this case.

```
<?xml version="1.0" encoding="UTF-8"?>
<!DOCTYPE document_root [ DTD definition details here can extend multiple lines ]>
```

The preferred approachin, my opinion, is to keep the XML file and the DTD file separate. A statement referencing the external DTD is placed after the XML declaration, but without the inserted DTD coding.

```
<?xml version="1.0" encoding="UTF-8"?>
<!DOCTYPE document_root SYSTEM "path/dtdName.dtd">
```

DTD Building Blocks

Document type definitions are primarily composed of variations on elements and attributes. DTD reserved words are always upper case; variables and other expressions may be lower or mixed case. Examples shown here do not encompass all the DTD features available for defining an XML document or file – only those typically needed for producing variable document content streams.

Common element statements & explanations:

<!ELEMENT elementName (#PCDATA)>	"PCDATA" stands for parseable data
<!ELEMENT elementName (#CDATA)>	"CDATA" means character data exempt from parsing
<!ELEMENT elementName EMPTY>	The element has no content or imbedded sub-elements
<!ELEMENT elementName (sequence of sub-elements)>	

Element qualifiers:

```
( )    Element or sub-element grouping
,      Element separator indicating specific order
|      Element "or" separator indicating choice
+      Element may occur 1 or more times
?      Element may occur 0 or 1 time
*      Element may occur 0 or more times
```

Common attribute statements:

```
<!ATTLIST elementName attributeName TYPE #KEYWORD>
<!ATTLIST elementName valueName (val1|val2|val3) "defaultVal">
<!ATTLIST elementName href CDATA #REQUIRED>
```

Attribute types and keywords:

CDATA	Means that the attribute contains character data
ID	The attribute provides a unique XML identifier name
#REQUIRED	An attribute value is mandatory
#IMPLIED	An attribute value is optional
#FIXED	The attribute has a fixed value

DTD Examples

Let's start by creating the DTD to define our XML sack lunch example in Listing 6-1 on page 38. Listing 6-2 (right) is the corresponding DTD. The best way to understand how the XML relates to the DTD is to compare listings. By the way, Listing 6-1 evaluates to "well formed" and "valid".

```
<?xml version="1.0" encoding="UTF-8"?>
<!-- Sack Lunch DTD for one or more weekdays -->
<!ELEMENT lunch (day)+>
<!ELEMENT day (sandwich, fruit, chips, drink) >
<!ELEMENT sandwich (#PCDATA)>
<!ELEMENT fruit (#PCDATA)>
<!ELEMENT chips (#PCDATA)>
<!ELEMENT drink (#PCDATA)>
<!ATTLIST day name CDATA #REQUIRED >
```

Listing 6-2. The "Sack Lunch" DTD

Now let's go on to another example – Listing 6-3 (below). We won't show the corresponding XML, but try to imagine a spreadsheet that calculates the value of on-hand inventory. The spreadsheet columns are item, description, price, quantity, and amount. There will be one row for each item.

```
<?xml version="1.0" encoding="UTF-8"?>
<!ELEMENT dataroot (row+)>
<!ELEMENT row (item, detail, price, quantity, amount) >
<!ELEMENT item (#PCDATA)>
<!ELEMENT detail (#PCDATA)>
<!ELEMENT price (#PCDATA)>
<!ELEMENT quantity (#PCDATA)>
<!ELEMENT amount (#PCDATA)>
```

Listing 6-3. The "On-hand Inventory Spreadsheet" DTD

Our last examples are Listing 6-4 and Listing 6-5 (both on the next page). They juxtapose a DTD (6-5) and a matching XML document (6-4). This XML document is actually an XML content stream generated from a database. This type of content stream would commonly be used in producing variable documents such as marketing mailers (an application that will be evaluated in Chapter 19).

While the content stream contains only two records (which we will call <row> in XML), any subsequent rows will be exactly the same in terms of XML tagging – only the content between tags will be different. The XML content stream is indented for readability, but would normally be aligned left for processing. The DTD essentially provides a blueprint and quality control for the content stream. As you might suspect, the content stream is well formed, and when compared with the DTD, it is a valid document.

```
<?xml version="1.0" encoding="UTF-8"?>
<!DOCTYPE dataroot SYSTEM "Mailer.dtd">
<dataroot>
        <row>
            <name>Anthony</name>
            <image href="file://Berries.jpg"/>
            <location>North Side</location>
            <offer>5% Discount</offer>
            <product>Paving Stone</product>
            <title>Mr.</title>
            <first>Anthony</first>
            <last>Able</last>
            <street>27 Able Street</street>
            <city>North Side</city>
            <state>NY</state>
            <zip>14533</zip>
            <code>14533-27200803</code>
        </row>
        <row>
            <name>Bill</name>
            <image href="file://Flowers.jpg"/>
            <location>South Side</location>
            <offer>5% Discount</offer>
            <product>Mulch</product>
            <title>Mr.</title>
            <first>William</first>
            <last>Ward</last>
            <street>27 Ward Avenue</street>
            <city>South Side</city>
            <state>NY</state>
            <zip>14536</zip>
            <code>14536-27200804</code>
        </row>
</dataroot>
```

Listing 6-4. A Variable Document XML Content Stream With Two Records (rows)

```
<?xml version="1.0" encoding="UTF-8"?>
<!-- DTD for Mailer -->
<!ELEMENT dataroot (row+)>
<!ELEMENT row (name, image, location, offer, prod-
uct, title, first, last, street, city, state, zip, code) >
<!ELEMENT name (#PCDATA)>
<!ELEMENT image EMPTY>
<!ELEMENT location (#PCDATA)>
<!ELEMENT offer (#PCDATA)>
<!ELEMENT product (#PCDATA)>
<!ELEMENT title (#PCDATA)>
<!ELEMENT first (#PCDATA)>
<!ELEMENT last (#PCDATA)>
<!ELEMENT street (#PCDATA)>
<!ELEMENT city (#PCDATA)>
<!ELEMENT state (#PCDATA)>
<!ELEMENT zip (#PCDATA)>
<!ELEMENT code (#PCDATA)>
<!ATTLIST image href CDATA #REQUIRED>
```

Listing 6-5. DTD for Variable Document XML Content Stream

Keep In Mind...

While using XML to create data or content streams, here are some considerations to remember:

- XML is hierarchical and extensible
- It describes content, not layout or formatting
- It can reference external objects
- Since content is separated from layout, content may be multi-purposed
- Well Formed: follows all XML conventions; Valid: agrees with the document definition
- Being "well formed" doesn't necessarily mean a document is "valid"
- There's no "right way" to code a document, but some approaches may be more effective than others
- Keep XML structures as simple and clean as possible; avoid unnecessary complexity
- White space is preserved; indenting sub-elements and inserting comments may help understandability, but may also interfere with integration processing
- Using fixed width fonts during XML preparation increases text control at the character level. Formatting XML documents is superfluous, distracting, and a potential source of problems. Plain text is best!
- Regarding plain text, a text editor is a better choice than a word processor, because in XML processing, "what you see is not necessarily what you think you have". The word processor is liable to introduce hidden characters that it needs to process the file.
- Since content structure and sequence drive processing, document design and content must correlate

Recommended Reading

Amiano, M., D'Cruz, C., Ethier, K., & Thomas, M. (2006). *XML problem-design-solution.* Indianapolis, IN: Wiley Publishing, Inc.

Stanek, W. (2002). *XML pocket consultant.* Redmond, WA: Microsoft Press.

Wagner, R. & Mansfield, R. (2003). *XML all-in-one desk reference for dummies.* NY: Wiley Publishing, Inc.

Supplemental Information

Organizations:

www.w3.org	World Wide Web Consortium
www.xml.org	XML news and resources site
www.podi.org	Print On Demand initiative
www.oasis-open.org	Organization for the Advancement of Structured Information Standards
www.apache.org	The Apache Foundation home page
cocoon.apache.org	The Apache Cocoon Project
Xmlgraphics.apache.org	Apache XML Graphics Project

Commercial Sites:

www.xml.com	O'Reilly XML site
www.oxygenxml.com	<oXygen/> XML Editor & Processor (Mac & Windows)

THREE: MECHANICS

In this unit we are going to expand our knowledge of the technologies we discussed in the prior unit: digital assets, data, and XML. Rather than talk about more features and capabilities for each, we are going to focus on managing and manipulating them. First we need to prepare our digital assets. Taking them into account, we next prepare our data and generate XML. After that, we clean up and massage the XML into a content stream, which we will load into our document layout.

Oh yes, the document layout! We are going to need one. So we have a fourth technology to look into – that's the layout design application. There are three things to consider: (1) designing the document layout, (2) loading the layout with the XML content stream, and (3) producing usable output in the form of a print stream.

7. DAM – Managing the Digital Assets

Moving from content definition to content management, the focus is on storage, security, and use. We take a manufacturing perspective: raw material, work in process, and finished goods.

a. Beyond Definition
b. Organization
c. Security
d. Space
e. Working With Images
f. Supplemental Reading

8. Desktop DBMS – Managing Source Data

In a similar vein, this chapter deals with data: storage, management, preparation, and conversion to XML for ultimate import into the document layout. However, there is still a little more work before we can actually execute a document load. On to the next chapter!

a. Data Acquisition
b. Basic Processing Patterns
c. Data Import
d. Data Enhancement
e. Content Stream Export
f. Supplemental Reading

9. Techniques – Managing XML Content Streams

Chapter 9 takes up where Chapter 8 left off – dealing with the exported data as an XML content stream. The chapter discusses three options for managing and making final preparations to the XML so it will be ready for import into thge layout.

a. Next Step
b. About The XML Content Stream
c. Using A Text Editor — What You Need To Know
d. Using XSLT — What You Need To Know
e. Using Perl Programming — What You Need To Know
f. Recommended Reading

10. Document Design: Integration & Print Stream

Now we need a document layout, a destination for our content. Chapter 10 tackles key variable document design and XML content issues to automate import and styling using Adobe InDesign. It then describes PDF generation and final print stream page sequence reorganization.

a. A Rough Sketch
b. Geometry
c. Page Masters
d. Styles
e. Print Stream
f. Recommended Reading

7. DAM – Managing Digital Assets

Discussion Points:

- ✓ Beyond Definition
- ✓ Organization
- ✓ Security
- ✓ Space
- ✓ Working With Images
- ✓ Supplemental Reading

Beyond Definition

Chapter 4 (Metadata About Digital Assets) discussed asset identification, description, and selection. This chapter will discuss issues associated with storing, protecting, and using digital assets. While this book is about digital printing and publishing, the extent digital technology has saturated the document production workflow from acquisition of content through creative design to digital output (computer to plate, digital printing, mobil device display, and Web publishing) makes this chapter and previously, Chapter 4, additionally applicable to static print production (for instance, offset lithography, gravure, and flexography).

Consider this thought: managing digital content for document publishing (variable or static) is much like performing a highly choreographed dance – perhaps a tango. Missteps mean toes get stepped on, dancers must recover their poise, clumsiness means extra footwork, and stylish artistry is lost.

The tango is a complex dance requiring both precision and flair. Bringing digital content together in an attractive and effective document also requires choreography and faithful performance. Figure 7-1, "Content Choreography", demonstrates the primary dance steps in bringing data and digital assets together.

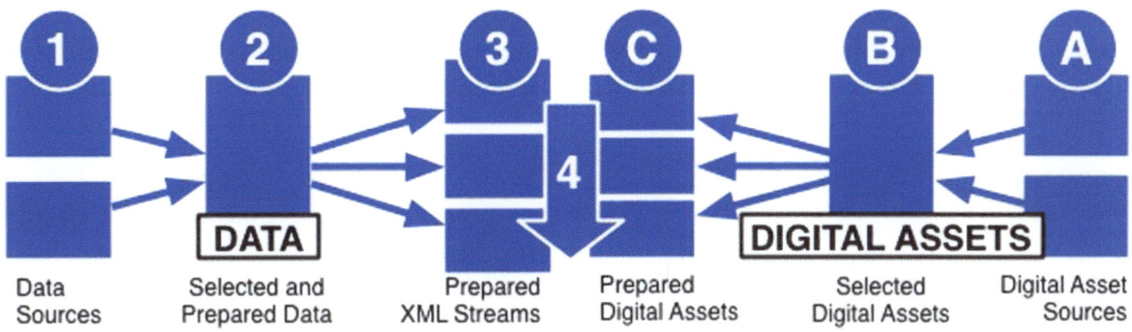

1. Data are selected from multiple sources
2. Data are merged, cleansed, analyzed, and enhanced
3. Data are ordered, grouped, and converted to XML

A. Digital and textual assets are selected
B. They are groomed and edited to meet requirements
C. They are grouped to match corresponding data

4. Coordinated XML and digital assets are ready for document generation, PDF conversion, and production

Figure 7-1. Content Choreography – A Dance Overview

Let's look at the dance partners' steps individually. Our focus here is strictly on the digital assets, although some of our discussion will resonate for data and text as well. We'll specifically discuss managing data and XML content streams in the next two chapters.

Organization

Structural Basics

Let's start with a look at basic issues about content organization and supporting practices. There is no "best" structure for managing content, but there is a "worst" one – no logical structure at all. Everything is in one place. You've probably know someone who keeps everything on their computer desktop. Granted, it's the personal preference of a single individual – as opposed to a married one (just kidding).

But seriously, there is an implied question: Does the storage arrangement facilitate effective and efficient management, retrieval, and recoverability of content, along with the frugal use of time and storage space? Here are some fundamental questions to consider when designing a new or evaluating existing storage schemes:

1. Will the storage scheme support the operations of the different organization entities?
2. Will the storage scheme support organization workflows?
3. Will the storage scheme simplify content classification and retrieval?
4. Will the storage scheme separate newly acquired or raw content from work-in-process, finished goods, and production archives, to protect a recovery path all the way back to the original content?
5. Will the storage scheme isolate newly acquired content until the files have been fully prepared: files named according to conventions, complete metadata recorded, key words selected from a controlled vocabulary, and appropriate retention determined?
6. Will work-in-process and finished goods be compartmentalized to insure that unfinished project elements are not accidentally run in production?
7. Will work-in-process, finished goods, and production archives be segmented by project or job?
8. Will projects or jobs be further subdivided into specific work components according to the department, work group, or person primarily responsible for completion?

There are no "right or wrong" answers to these questions (except perhaps "no!"). Any design evaluation needs to take them into account. Ignoring them risks potential missteps, and compromises future ability to recover from mistakes. Failure to adequately control use of digital content threatens end-product quality, customer satisfaction, profitability, and ultimately organizational viability.

Rudimentary Flows

Moving from the dance floor to the shop floor, let's look at how content circulates through the document development process. The flow is very much like a manufacturing process in a factory, except that our material is primarily composed of intangibles – various kinds of digital graphics. By the way, text and data may be considered another form of digital asset, which has a similar flow (as shown in Chapter 8).

Figure 7-2 (next page) illustrates the marshaling and movement of content through the major stages of content storage and document development: raw materials, work in process, and finished goods (plus scheduled backups for each area).

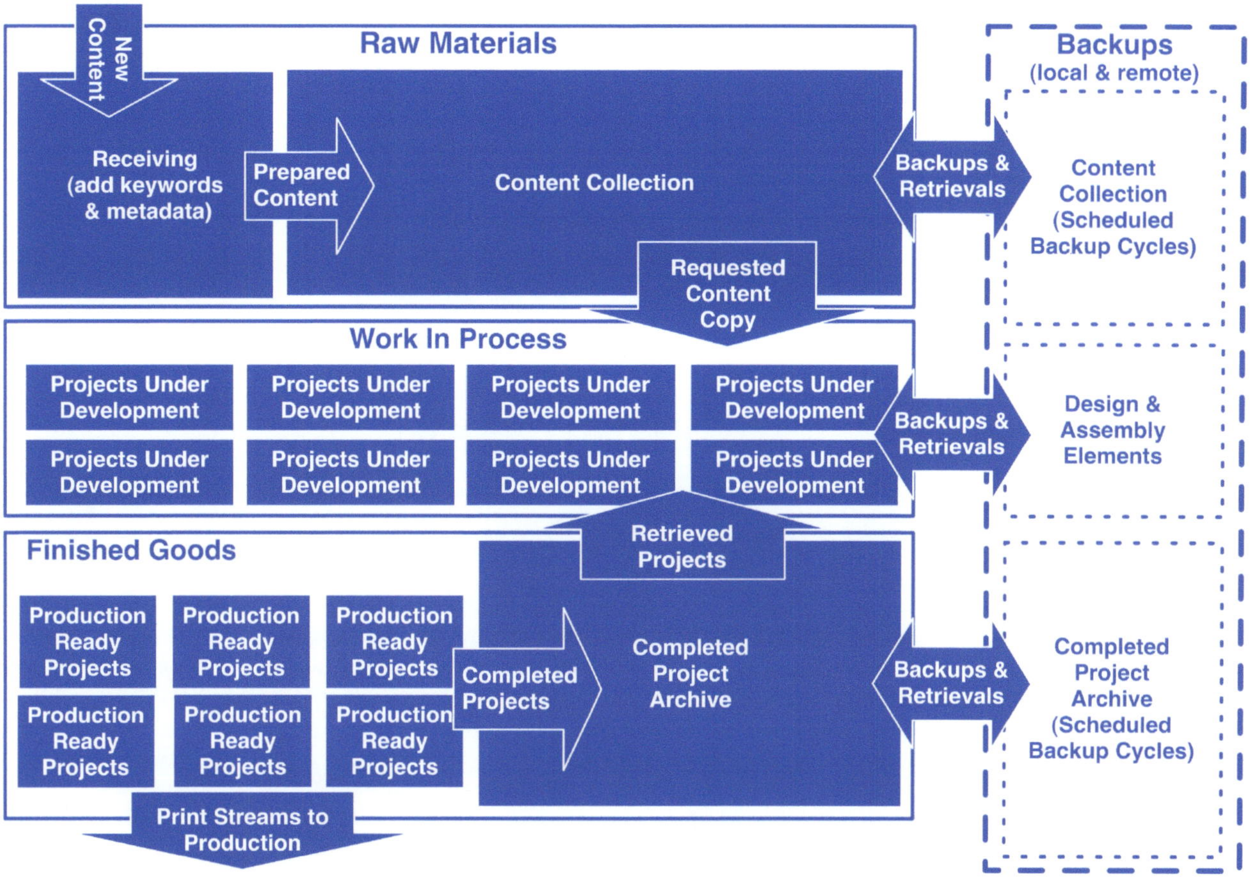

Figure 7-2. Content Storage Partitioning and Circulation

Raw Materials

Newly acquired and customer-supplied content enters temporary receiving storage for initial preparation. The digital graphics are evaluated, existing metadata supplemented where needed, file naming conventions applied, and a decision made on a continuing storage location. Note that the digital graphic itself is not altered, but kept "as received".

The prepared digital assets are moved from receiving and logged into the content collection. Developers or requesters anticipating use of this content are notified. The content is placed in the collection on either a semi-permanent basis or a project basis. Project-basis content (usually supplied by the customer) is retained for the duration of the project or according to final disposition instructions from the customer. The content collection is considered as semi-permanent in that obsolete content or digital assets with expired licensing are periodically weeded out.

Work In Process

Required assets are copied (not moved) into project storage from the content collection. In project storage, images are converted to the standard working color space, corrected, enhanced if needed, and sized according to requirements. After work-up, graphics are placed in a project staging folder for integration into the XML content stream, and ultimately into the document layout. After successful proofing, the design software generates a PDF print stream and production package for the project. (*We will elaborate on work in process in later chapters.*)

Finished Goods

The entire project is then moved to finished goods. The print stream and production package are copied to production staging for print or other digital generation. The project (all project artifacts) is moved to the project archive, where a decision is made regarding term of storage:

- Temporary – return to the customer or dispose after a set amount of time
- Short Term – retain for follow-on work
- Long Term – compress and retain for reference or potential reuse at a much later time

Security

Permissions

We're not talking about system and network security, although obviously related. We're addressing the security of the digital asset management system (DAMS) and its content. Whether raw content (originals and newly acquired assets), content being used in development or production, completed work, or project archives. The assets represent a considerable investment in funds, time, labor, future opportunity for reuse, and customer good will. They have significant value and deserve protection.

What we are going to talk about is who can access what DAMS resources (folders and folder content) and what actions are permitted on specific accessed items. This does not have to be complicated, but there are two caveats: First, the security arrangement must support (not obstruct) the operations and workflows within and between working groups in the organization. And second, there is no ideal combination or scheme that will meet the needs of every organization – security plans need to be customized – tailored to the organization.

The example below demonstrates one possibility, and is only intended to stimulate thinking. It is by no stretch of the imagination "ideal". Its provisions are certainly debatable, but it may provide a starting point.

1. DAMS system level:
   a. System Owner – full access, full permissions, all levels
      i. Activities usually limited to periodic inspections and exception conditions (operational issues and emergencies). See DAMS Administrator details below.
      ii. Day-to-day DAM administration delegated to the DAMS Administrator
   b. DAMS Administrator – full access, full permissions, all levels
      i. DAMS System maintenance – problem resolution, optimization improvements, application of version updates and patches, scheduling backups, and performing recoveries.
      ii. Grant and revoke DAMS logons and passwords
      iii. Grant and revoke permissions
      iv. Permissions: Create, Read, Update, Copy, Move, and Delete folders and content
   c. DAMS Librarian – full access, limited permissions, all levels
      i. Folder Permissions: Read, Update, Copy, and Move
      ii. Subfolder Permissions: Create, Read, Update, Copy, Move, and Delete
      iii. Content Permissions: Create, Read, Copy, and Move
   d. DAMS users – specified access, no system level permissions

2. Folder level:
    a. Folder Owner – specified folder and content access, limited folder permissions
        i. Folder Permissions: Read, Update, and Copy
        ii. Subfolder Permissions: Create, Read, Update, Copy, Move, and Delete
        iii. Content Permissions: Create, Read, Update, Copy, Move, and Delete
    b. Folder User – Specified folder access, specified folder permissions
        i. Folder Permissions: Read
        ii. Subfolder Permissions: Read, Update, and Copy
        iii. Content Permissions: Create, Read, Update, and Copy
3. Content level (content or subfolders within the parent folder):
    a. Content owner – content access within specified folder, limited permissions
        i. Content Permissions: Create, Read, Update, Copy, Move, and Delete
    b. Content user – specified content access, specified permissions
        i. Content Permissions: Read, Update, and Copy
    c. Limited Content user – specified content access, Read Only permission

Usage Control

What we want to briefly talk about is the prevention of missteps such as the accidental loss of work on a digital asset through untimely replacements or accidental overwrites. This issue comes to the forefront where multiple people share responsibility for working with the same asset. Simply limiting responsibility to one person isn't necessarily the most productive approach, nor does it insure against blunders.

Standard operating system functionality and metadata (PC or Mac) can provide a partial audit trail of file activity. SOP's (standard operating procedures) promote recording additional information manually. And of course, backup and recovery systems are available. But who wants to go to the trouble of frequently recovering files? What if a file wasn't backed up before it was overwritten?

A preventative measure in software application development shops is a version control application. Adobe Software includes one called, " Version Cue", in their "Creative Suite" graphics toolkit. It offers file "check out" and "check in" capabilities similar to how a library functions: when a book is signed out, nobody else can read it until it's returned.

Backup

Turning to backup, it should be done frequently. System level backups need to be done daily. Many shops do weekly and monthly backups as well. Some use a three-step backup cycle the most recent backup is ultimately rotated out after two more cycles.

An organization could ask: "Who has the time? Who has the money? Who has the space?" The answer, of course, is another question: If your digital assets are your organization's "family jewels" (so to speak), how long can you stay in operation without them? Then following up, can you afford to replace them (if even possible), and how long would that take? If they are irreplaceable (chances are likely!), then what?

Perhaps a little over dramatic, but even in terms of a single project, a loss could represent overtime labor costs, wasted materials and supplies, missed schedules, and unhappy customers. Enough recurrences of that sort and you're staring at the face of a failing or already failed business.

Space

Space – not referring to the "final frontier" though our topic does seem somewhat expansive. Wordplay aside, the question is, how much storage capacity will DAM require. To be blunt, there is no way to tell at this distance. While there may be similarities among enterprises, each is unique in terms of size, scale of operations, infrastructures, quantities of digital assets, organization, work flows, and products produced. Consequently, requirements will differ to a greater or lesser degree, meaning that each DAM installation will also be unique in some sense.

DAMS vendors have the experience and formulas that help in estimating storage requirements. However, below is a rough (very rough!) recipe for deriving a rather dubious approximation (guess) of the amount of space that might be needed.

The rationale for the sequence of calculations is that a percentage of each area influences the areas that follow. A percentage of content collection adds to the other kinds of content-related files, which are presumably being worked upon. This gives work-in-process. A percentage of work-in-process influences finished-goods; finished-goods contains all archived projects, which is a further reflection of the content collection. The content collection exists because it has been created by doing and for doing projects. As a rule-of-thumb, we will always round calculation results upward.

Space Requirements "Guesstimate"

1. Total all your graphic images by quantity and byte size, giving your content collection.
2. Total all other related digital assets (text, data, unpopulated and populated document layouts, content streams, and print streams) by quantity and byte size.
3. Take the step #2 totals, add 25% of the step #1 results, then multiply these by 1.5, giving work-in-process .
4. Take 50% of step #3 and add the results from step #1, giving your finished-goods-inventory.
5. Add together your results from step #1, step #3, and step #4, giving your total operating environment.
6. Double the results from step #5, giving your total estimated environment requirement (including backup).

As you can tell, this is an extremely rough analysis. There is no way of knowing whether the final result is high or low, but it does provide food for thought. Note that the estimate does not include application software, operating system software, or internal business files. These can be "guesstimated" similarly. Percentages and multipliers can also be tweaked up or down. In other words, you can feel encouraged to play with the numbers – model it in a spreadsheet.

For those who get anxiety attacks from such calculations, a statistical approach may provide more comfortable insights. For those who have a penchant for absolute accuracy, abhor risk, and rarely feel secure with "field expedients", I'm sure there is estimating software available for a price.

However, as casual or non-scientific it might appear, In terms of invested time, this is not a meaningless exercise. Whether there is any value in the calculations and final results, the inventory analysis (byte and file counts) provides a basic starting data for any estimating approach.

Working With Images

The original condition of the digital images is protected in the DAM collection. "As-modified" images are stored in the finished goods archive. For use in a project, digital assets are selected and copied into the appropriate project space allocated in working storage. Figure 7-3 encapsulates the basic steps in image preparation.

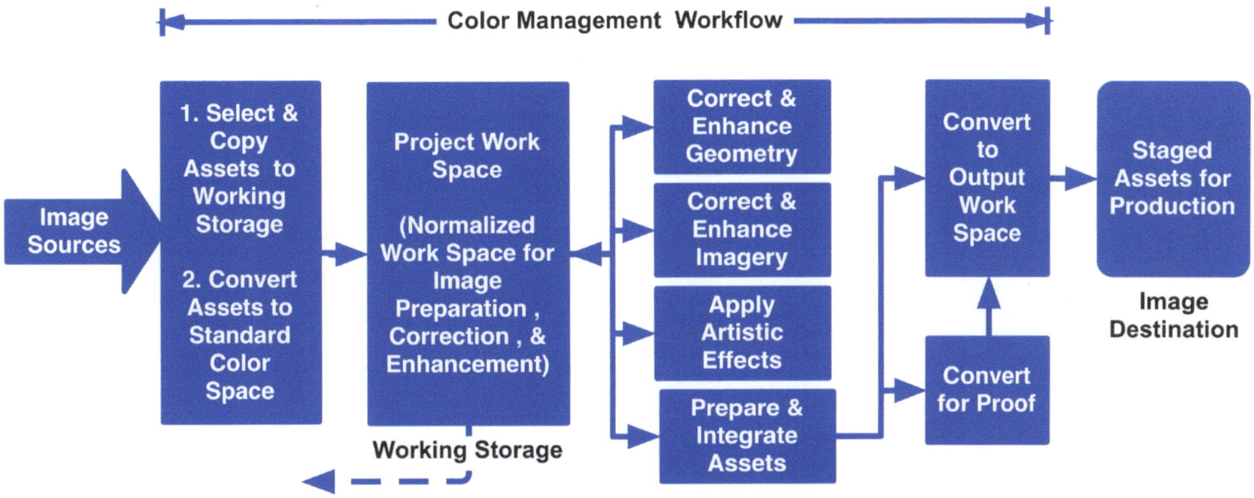

Figure 7-3. Image Preparation Workflow

After selected digital assets are moved into working storage, what has to be done and why? The next two subtopics explain this. It seems that no two experts have the same list of issues or ways to address the issues. While there is overlap, some points get left out. The lists presented in the following two subtopics are an edited and abreviated compilation from two very comprehensive books on digital imaging (Baxes, 1994, and Russ, 2002). Additional perspective came from a book on color management by Fraser & Bunting (2005). More recent versions of these three references are probably available, if you're wondering about about the material being dated. You can take confort in the fact that issues are very basic – not cutting edge.

Potential Image Problems

Issues posed by digital content may have a variety of causes that usually relate to embedded color profiles, embedded fonts, copyrighted fonts and images, file format, file size, and assumptions about appropriate color space – particularly in regard to the generation of Adobe PDF (Portable Document Format) files. More complex problems, which generally occur at capture (according to Baxes, 1994, and Russ, 2002), directly relate to image quality as opposed to any peripheral factors:

- Image distortion
  - Barrel distortion (outward distortion from the center)
  - Pincushion distortion (inward distortion toward the center)
  - Vignetting (edges and corners darker than center)
  - Chromatic aberrations (color fringes on object edges)
- Image resolution
  - Spatial density and related characteristics

- Brightness
- Contrast
- Color quality
- White balance
- Detail resolution
- Tonal quality and range
- Noise (luminance, color, compression, digital artifacts and patterns)
- Out-of-focus (specific objects or overall)
- Motion blur

- Image composition
- Viewing geometry
  - Image dimensions
  - Viewing distance
  - Aspect ratio (rectangularity)
- Compression issues (loss of image quality from high compression ratios)

Image Repair & Preparation

Repairing damaged images or problems in scanned and photographic images can be difficult. In some cases it may be simpler to replace the image or find a substitute. Repair can involve a number of activities:

- Correct original photographic conditions or capture errors.
- Adjust white balance.
- Adjust overall contrast or tonal range.
- Remove color casts.
- Remove motion blur and correct out-of-focus issues.
- Adjust color and tone in specific objects or areas to bring out highlights, mid-tones, shadows, and desaturated colors.
- Sharpen overall image focus.
- Enhance edges and object contrast.
- Remove noise.

Images that have already been processed or modified as a result of artistic work may still require some corrective action to make them more usable or able to achieve desired results in production. Corrective actions may include the following:

- Address artistic filters – change the filter, remove the filter if possible, or replace the image.
- Correct image geometry (rotation, scale, warping).
- Segment image (remove distracting or useless information).
- Shrink, expand, or mask selected objects to emphasize objects of interest.
- Crop and rotate image to final size and orientation.

- Flatten multi-layer images.
- Flatten transparencies.
- Insure fonts are embedded.
- Convert images to correct destination color profiles.
- Insure correct profiles are embedded in images.
- Convert images to PDF format (if needed), allowing for appropriate color space and image rendition based on production destinations.

Supplemental Reading

This chapter is obviously not an exhaustive explanation of content management. It is limited to information and practices primarily aimed at facilitating VCP or variable content publishing and printing. The following text provides a very complete discussion of content management and content management systems.

Boiko, B. (2005). Content management bible, 2nd edition. Indianapolis, IA: Wiley Publishing, Inc.

These next two books provide very comprehensive information about working with digital images. This information is not only helpful for image use, but also provides important considerations when evaluating images for acquisition, collection, and potential application. Newer editions may be available.

Baxes, G. (1994). Digital image processing principles and applications. New York: John Wiley & Sons, Inc.

Russ, J. (2002). The image processing handbook, fourth edition. Boca Raton, FL: CRC Press LLC.

This last book presents a fine discussion on color management, which is a crucial element in all aspects of image preparation and use in digital and printed documents.

Fraser, B., Murphy, C., & Bunting, F. (2005). Real world color management, second edition. Berkeley, CA: Peachpit Press.

In addition to these suggestions, it may be very helpful to familiaize yourself with Adobe Illustrator for vector graphics and Adobe Photoshop for raster graphics. Both applications come with their own "help" i9nformation and tutorials. Lastly, if you have a digital camera, it will be helpful to work on image capture in light of the issues listed above. There are a number of ways to capture digital images; the digital camera is only one alternative.

8. Desktop DBMS – Managing Source Data

Discussion Points:

- ✓ Data Acquisition
- ✓ Basic Processing Patterns
- ✓ Data Import
- ✓ Data Enhancement
- ✓ Content Stream Export
- ✓ Supplemental Reading

Data Acquisition

We continue to deal with process techniques and technologies, but we are getting closer to document design, integration, and print stream generation. We are increasingly building on prior chapters. You have been hearing previously mentioned terms, and you will begin seeing diagrams with similarities to earlier figures.

At point of acquisition, data may come from multiple sources and multiple data sets, both presenting consolidation issues. For example, The data may already exist in an organization's accounting, marketing, or other information systems. Other data may come from external organizations or agencies, intended as either supplements or primary data.

According to Rud (2001) and Westfall & Blaxton (1998) writing on data mining, and Mattison (1999) writing on data warehousing and knowledge management, in-house and incoming data may present the following problems:

- Multiple data structures and schemas
- Multiple formats and encoding
- Multiple platforms and operating systems
- Multiple security levels
- Multiple originating or transmitting locations
- Multiple transport media
- Multiple spoken languages

The data will be brought into the document development environment in much the same way that digital assets were brought in for storage and preparation. In fact, Figure 8-1 (next page) should look very familiar; it's essentially Figure 7-2 with minor variations.

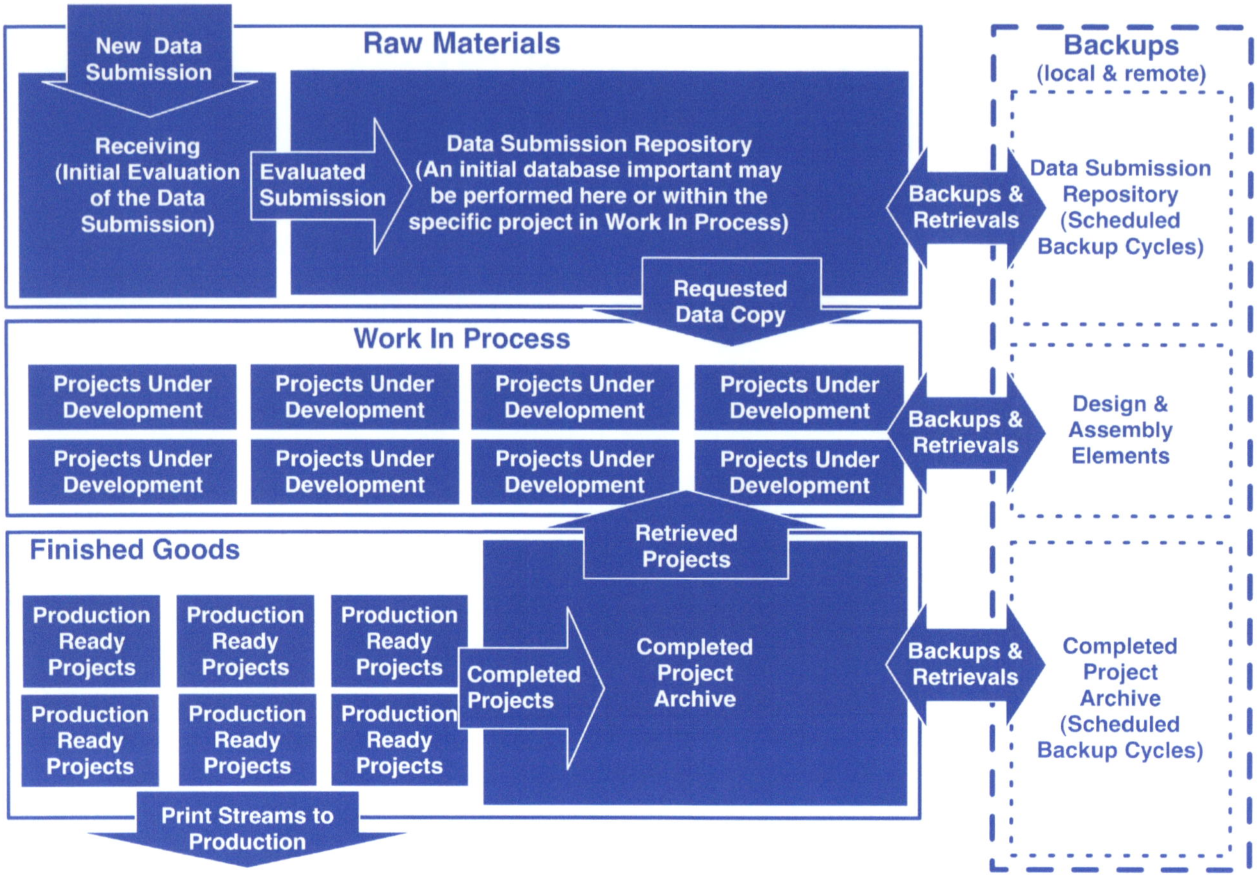

Figure 8-1. Data Acquisition, Preparation, and Storage.

Acquisition processes involve more than initially obtaining data. They include its storage, administration, and the necessity to provide data security and recoverability. Management of data within the development environment is based on the same logic that was applied to digital graphic content. The primary difference at an environmental level is that acquired data may be converted into a more workable format in the "raw materials" stage. The early conversion decision will probably be made as a result of an initial evaluation in "Receiving" and in consultation with the designated project team. Let's first build on our discussion of data processing structures from Chapter 5.

Basic Processing Patterns

Now that we have the data, how do we prepare it? To determine that, we need to look at basic data structures from a processing perspective. Preparing data for integration into a design layout involves more than changing field values, spelling, or character case, but also includes making structural changes to the content container or file layout.

There is a high probability that acquired data will take the form of a sequential file with some sort of delimiters (for instance commas or tabs), a spreadsheet, or set of these file types. Keeping it simple is an important objective; one approach may be more advantageous than another. Here are the two primary processing alternatives available: sequential processing and table processing (spreadsheet or database).

Sequential Processing

The easiest part of sequential processing is the repetitive program loop that steps through a file from start to finish one record at a time. This may involve only reading each record, or it may also include making selected changes that require concurrently reading and potentially modifying each record (Figure 8-2).

The simplicity of a relatively plain read-write loop quickly dissipates with the addition of field selection. Looking at Figure 8-2, though a record is said to contain fields, in actuality it may be nothing more than a string of characters terminated by a line feed (LF), carriage return (CR), or both (CRLF). Difficulty comes when determining the individual fields within each record. One approach is calculating substrings.

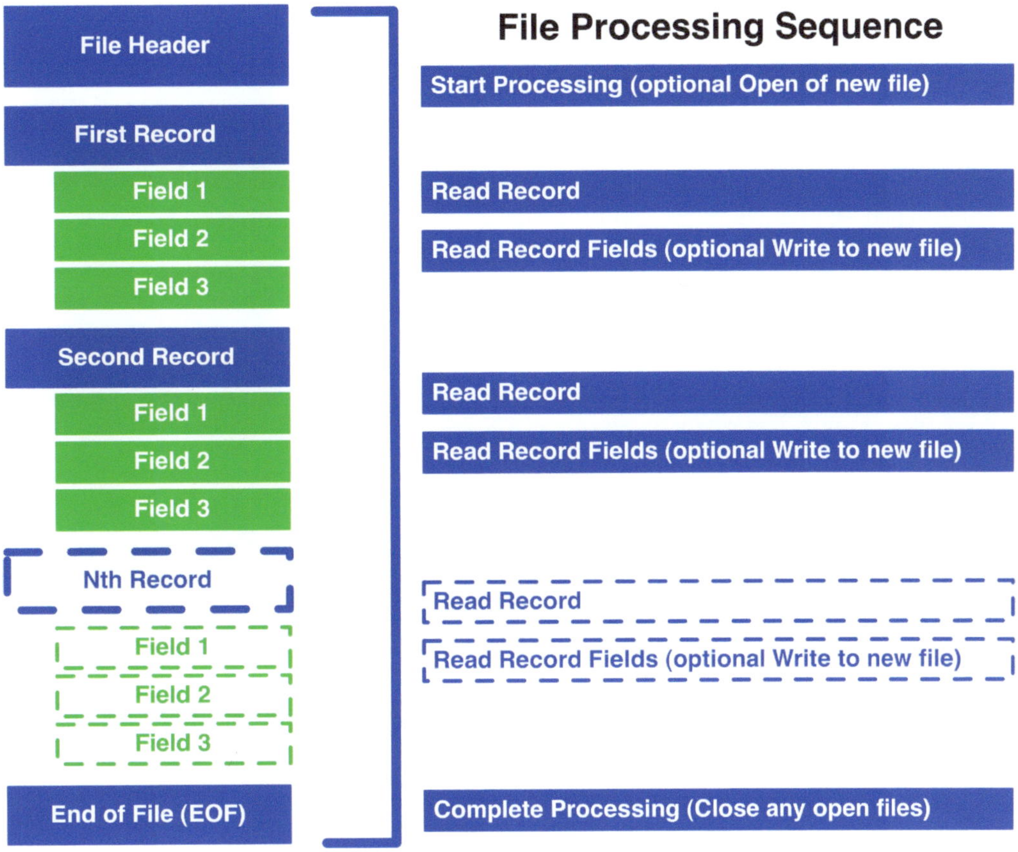

Figure 8-2. Sequential File Processing Model

For example, a text file may contain a set of records, each record 26 characters in length plus line feed. Below is the character string and a description of the logic involved:

"abcdefghijklmnopqrstuvwxyzLF"

1. Characters in each record string occupy a position numbered sequentially from 0 through 25.
2. If the first field were six characters, the substring would start at zero (the first character position) and count six character positions (zero through five), giving "abcdef".
3. The next field (substring) would start at position six, and suppose it is ten characters long (six through fifteen), which gives "ghijklmnop".
4. The 26 character string in every record in this file would be divided into substrings (fields), each adhering to a designated starting character position and number of characters (fixed record and field lengths).

As long as every record (every string of characters) adheres to the same designated substring pattern, everything is fine. However, there is no assurance that deviations won't exist in the file. To counter this, we need substantially more file evaluation and programming effort – more work and more expended time. An alternative approach requires that data files have delimiters. A common technique is the use of comma-separated values (CSV's) to identify record fields. Check out this example:

"John Smith , George Lipton , Mary BentonLF"

A problem can occur if some delimited field has text containing punctuation (i.e. a comma) that is also used as delimiters. The embedded punctuation confuses the processing program, because what is normally a single delimited field now appears as two fields. Since our hypothetical program was written to deal with three names per record, it now has a record that seemingly has four names (highlighted in last name first order):

"John Smith , Lipton, George , Mary BentonLF"

The best way to solve this problem without more programming is to use tab-separated values. Using tabs to separate character groupings significantly reduces the probability of delimiter errors. The tab-delimited example below (tab spaces highlighted) properly clarrifies the character string for program processing. The comma is treated as just another character, maintaining record integrity.

"John Smith	Lipton, George	Mary BentonLF"

Another complications is the prospect of merging two or more sequential files. In addition to the mechanics of identifying individual fields in each record, there is the need to build a merge process. The merge process may not be so simple, because there is the necessity of addressing any relational aspects between the individual source files. Figure 8-3 provides a picture of the considerations involved.

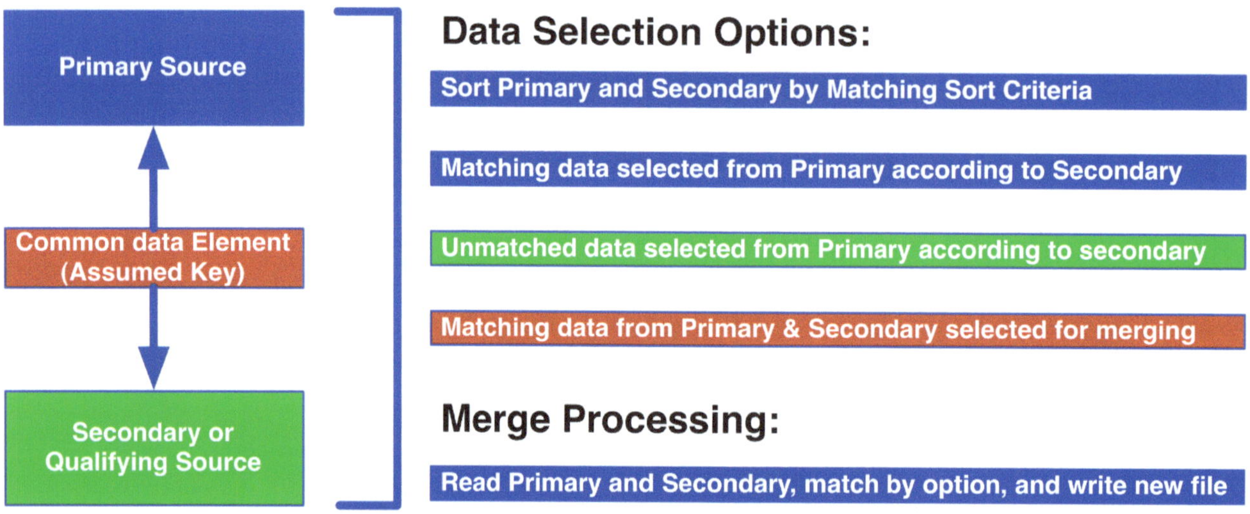

Figure 8-3. Sequential File Merge Processing

Consolidation approaches may join matched records from both data sets; use all records from one data set, updating particular record fields from matching records in a secondary data set; or only use records from one set where no matching records are found in the other data set. Combinations or sequences of these processing approaches may be needed, depending on project data requirements.

Perhaps the most significant issue is the degree of probability that records from the respective files will consistently match (or not match), taking into account the number of records per file and the possibility that individual files may come from different systems and different organizations. The easiest way to resolve this potential problem is to avoid it by using different tools and techniques.

Table Processing

Why not just get out of all the sequential programming and processing, and just use a spreadsheet? That would certainly be a step in the right direction. After all, spreadsheets can usually import delimited files with little or no preparation. Displayed import results can be visually examined. Problems within columns can be identified and rectified.

But what happens when files (now spreadsheets) have to be merged? We could be faced with more work, copying and pasting whole columns or writing program logic – how time-consuming or accurate will that be? Granted, spreadsheets are excellent for manipulating data, but why use a number cruncher to manipulate potentially relational data sets?

Why not simply use an application designed to display data in table views, plus the capability to inter-relate disparate data tables? Figure 8-4 illustrates some essential VCP data preparation processes that imply relational functionality beyond spreadsheet capabilities. The relational desktop database has a complete compliment of tools (including point-and-click scripting) for managing and preparing data for VCP.

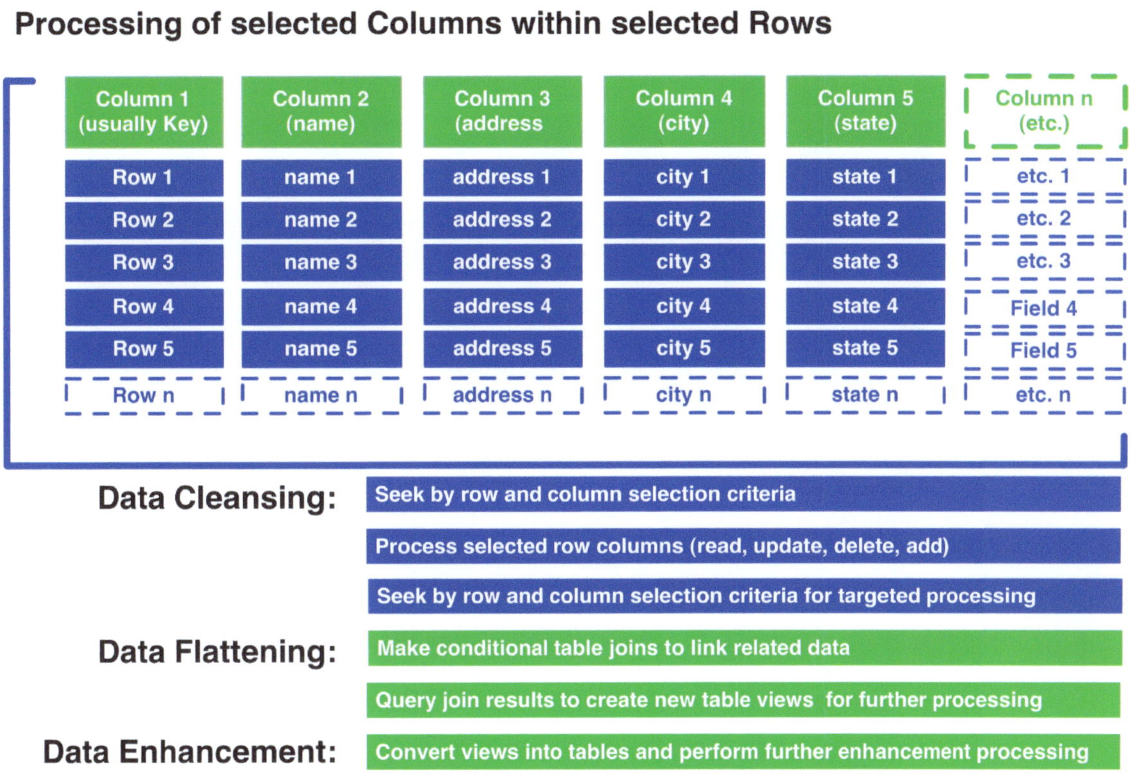

Figure 8-4. Database Table Processing Model

Based on the processing issues we've covered so far, we'll step through a data preparation process from start to finish. Figure 8-5 (next page) portrays the basic preparation sequence by the numbers. Then, after surveying issues we may encounter during processing, let's follow all this up with an example using a desktop DBMS (FileMaker Pro) as our primary tool.

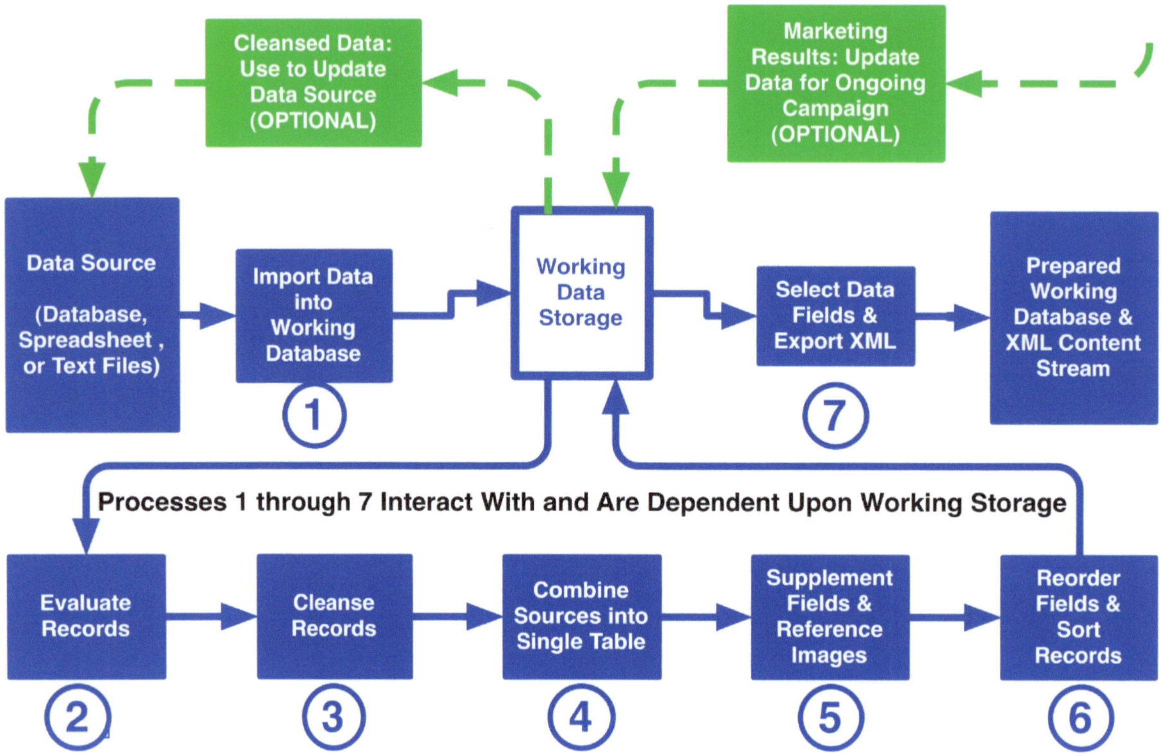

Figure 8-5. Data Preparation Workflow.

Data Import

The primary data preparation objective is to take acquired data and convert it into a single stream of clean data that has been appropriately enhanced, organized for automated integration into a document layout, and converted to XML. The first step is bring the raw data set (or sets) into an application software environment that will simplify data preparation – the desktop DBMS.

While we will be using FileMaker Pro throughout the data preparation discussion, this doesn't imply that it is the only or the best alternative. If Microsoft Access were available on a Mac, it might be the center of our discussion. However, FileMaker does a very clean conversion from tabular data into XML. We'll cover this at latter.

Importing raw data from some type of delimited file or spreadsheet is relatively easy. Some minor setup may be needed for delimited text files or spreadsheets, and import processing may require establishing receiving database tables before the actual import process:

1. Check the text file or spreadsheet to determine whether the first row contains data or column headings.
   a. If the first row contains column headings, note them for possible use in the database import setup.
   b. If the first row contains data, use the text editor or spreadsheet application to insert a new first row with simple descriptive column names (for instance on address data, "first, last, street, city, state, zip").
2. The database application import processing setup may only require the file path and file name, file type, delimiter type, and whether the first file row contains headings (life is much easier if it does).
   a. As an additional setup requirement, some database applications may require creating an empty table as the destination for the imported data.

b. If so, the application may also require pairing each source data column heading with a corresponding destination table column heading (this pairing may be done automatically depending on the DBMS).

The one word of caution: Some care and attention to detail should be exercised, especially if constructing an empty destination table for receiving imported data.

- The first column of the new table should be an automatically incremented row number (if that application feature is available, the values will be generated during import).
- Most fields for imported data (including zip codes and static numbers) can be designated as text.
- Fields for quantities and monetary values that may be used in calculations should be number fields (some databases may require explicit designations such as integer, floating point, or fixed-place decimal).

Data Enhancement

When we say data enhancement, we are talking about a lot more than just error correction and cleaning up "dirty" data. We are talking about a series of steps that will turn raw data into a stream of content that can be used repeatedly to produce a variety of variable data documents.

The best place to start is by getting a basic understanding of how and why data becomes sullied. People who deal with knowledge management and data mining have become experts in this topic – it's been an issue for a very long time. Let's create a list of issues and corrective measures. Westfall & Blaxton (1998), Mattison (1999), and Rud (2001) are good references. There isn't a single comprehensive list. Everyone has different opinions and ideas, so we've removed duplications, corrected omissions, and pieced together our own list.

Design & Programmatic Problems

Some problems trace back to structural and functional mistakes or omissions. Other problems are related to the programs that interact with the databases or files – faulty application design and errors in program code. Here are a few of the issues:

- Normalization – inadequate or omitted normalization steps during database design, resulting in:
  - Data redundancy – repeating or duplicate data elements between individual tables
  - Inconsistent multiple values for the same data element – failure to establish lookup tables
  - Duplicate records – unique record key not established or enforced for key-related content
- Referential integrity between parent and child tables is a design problem exacerbated by processing
  - Update anomalies – updating the wrong parent or child record
  - Insertion anomalies – adding a new record into the database using an existing key
  - Deletion anomalies – deleting a parent record and orphaning its child records or removing the parent record while deleting a child record
- Data integrity – database design deficiencies or programming logic that degrades the value or accuracy of the data, resulting in the following problems:
  - Incomplete data – failure to select, capture, or account for all records or data elements within records
  - Lost data – faulty program procedures that drop data elements or whole records during processing
  - Missing data – data elements that were never comprehended during database analysis and design
  - Corrupted data – data elements or whole records made unintelligible or unusable during processing

People-related Problems

Some problems stem from managers looking for quick solutions to store information beyond available data elements. Others may result from careless or imprecise data entry (clerks, bookkeepers, customer service representatives, and sales people). Problems may directly relate to lack of quality assurance processes and inattentive quality assurance analysts. Here are some of the problems:

- Incorrect or inconsistent spelling, and transposed characters
- Ambiguous names or variations on names
- Incorrect, inconsistent use of capitalization, upper, and lower case
- Improperly merged fields or merged content within a field
- Extraneous characters or leading and trailing blank characters
- Empty or null values, excluded or omitted content, and incorrect, irrelevant, or unknown content
- Abstract or meaningless coded values, incorrectly selected codes, and unit conversion errors (pounds to dollars)
- Cross-record duplication or data entry (data for separate records erroneously mixed)
- Under-utilized databases (spotty data and partially completed records) or excessively large databases (under-maintained and out of date records)

Cleanse

Before effective use, data needs cleansing. This activity may require significant amounts of analysis, data manipulation, and time. Complete problem detection and correction may not be possible.

Problems that reflect source data issues are going to be difficult to repair in a data extract since it no longer part of the original body of data. Missing record or field content without source data access for research may defy repair. Missing records may not even be detectable – especially if control counts based on the source data are not provided.

The first step in cleansing is triage (analysis) of the data set to determine corrective measures:

1. Based on project objectives and requirements, what types of record problems are irrelevant and can be ignored, what types of problems should be resolved, and what types may be irreparable?
2. Based on the ratio of detected problem records to the total number of records in the data set, can the data set be sufficiently cleansed or should it be replaced (assuming that's even possible)?
    a. Is the replacement likely to be any better?
    b. If actually replaced, what is the condition of the replacement?
3. Regarding individual records, on what basis should a record be discarded or replaced?
    a. If the question is about record discard, how significant is the record in relation to the entire data set?
    b. If the question is about replacement, what is the likelihood that the replacement will be any better?

Depending on findings, the cleanup process can start. The effort is aimed at not only resolving issues identified during analysis, but also the simplifying the dataset down to the bare essentials. Here are some common corrective measures:

- Make sure field names are accurate
- If possible, resolve invalid, incorrect, or missing content, including four-digit zip code extensions

- Check and correct spelling, inconsistent spelling, or transpositions
- Correct upper and lower case issues
- Eliminate duplicate records (outright duplications or duplicates with some conflicting values)
- Resolve ambiguous names or name variations
- Resolve cross-record duplication (if workable, set up "households")
- Make appropriate unit conversions and decode or correct coded data values (make values meaningful)
- Remove extraneous characters, leading and trailing blanks, unintelligible, meaningless, or irrelevant data
- As a final step, make certain that all irrelevant or irreparable records and data elements are removed

Combine

There's nothing like stating the obvious: combining is only relevant where two or more data sets are being used to provide additional primary data or supplementary data relating in some way to the primary data.. They may be from the same or different data sources. The first question to arise is whether to combine before cleansing or after. It is best to first resolve as many issues as possible. Combining "dirty" data may create more issues and more work rather than less. However, there may be cases where the process of combining by update resolves missing or corrupted data elements. This underscores the need for careful preliminary data analysis.

There are three ways to combine data sets: concatenation, update, and merge. The simplest one is concatenating multiple data sets into a larger set of more records. The concern in this process is to make sure there are no overlapping records in both sets of data. Possible duplicate records are easy to detect; simply sort both data sets into the same order and match them. Records that match are probably duplicate; one may be discarded. However, prior to discarding, the assumed duplicates need to be validated:

- Are the records really duplicates or is there an error in one of the matched fields?
- Do they actually contain the exact same information?
- Does one record contain more or cleaner information than the other?

The second method of combining data is by updating a primary data set with data from a secondary file. An example would be updating incorrect, missing, or omitted data in the first extract with a second abbreviated extract from the same source data. Again, confirm that the two data sets match before physically executing the update.

The third method – a merge – combines selected fields from two matching but different records in different data sets into a new set of records in a single new data set. An example of this would be combining descriptive and behavioral information from two sources to produce a single set of data.

When setting up and executing such a merge, it is important to maintain data source identity in order to assure data integrity, traceability, and recoverability:

- Record the primary keys from both records in the new record
- Record in the new record the respective matching fields, if other than the primary keys
- Create a new unique primary key for the new record
- Group fields in the new record according to their respective source

Supplement & Sort

The next data preparation step is to supplement the original data records with added fields based on existing record content. The additional record elements serve two purposes:

1. To address business logic at a database structural level by incorporating it into the record level structure as a more readily accessible field element.
    a. This eliminates the need to process extraneous logic during variable document generation
    b. It provides a reviewable, revisable audit trail prior to document generation, if corrections or changes are identified
    c. And it provides a basis for updating the working database to accommodate new requirements as a campaign progresses or a new one is launched
2. To align the working database (and ultimately the XML content stream) to satisfy document design and processing requirements.

The types of additions to data records are a limited variety, and easily made with our desktop DBMS:

- Insert processing flags to facilitate database-level decision making:
    - Binary flags that are either turned "on" or "off"
    - Optional values that are set from a controlled list
    - Relational values or thresholds (equality, non-equality, greater than, and less than)
- Add a new field in which to concatenate the content from multiple fields into a single field to simplify content stream processing
- Replicate a field as many times as it is used in the document design, if used more than once
- Summarize quantitative fields (amounts and counts) if a total is needed
- Insert a field to reference the location (URI or storage address) of each digital image used in the document design that is dependent on other data in the record.

The final step in the preparation process is simple – sort the entire set of records into zip code order to take advantage of postal bundling rates. Everything is now ready for export to an XML content stream. However, before continuing, let's walk through a short example to demonstrate the logic and simplicity of initially preparing a data stream.

"Friends of Dogs and Cats"

The rescue organization, Friends of Dogs and Cats, is running a charity campaign using a marketing mailer to raise more money for growing needs. A second purpose in the campaign is to find more homes for orphaned pets.

The mailer will be a variable document, personalized, in full color, and sent to people who have adopted dogs and cats from the organization in the past. A special discount will be offered for new adoptions if accompanied by the mailer. The issue of multiple adoptions by a single person or family was resolved during data cleansing. The organization has kept a record of all adoptions, including a digital photo of each pet. This will provide the source data for the campaign.

The Data

The source data set for the campaign has already been imported into a database and cleansed. Everything was on a single delimited text file, so combining was not an issue. (*They plan at a later date to go to the county to get a list of licensed pets, and combining data will be needed then.*) Listing 8-1 shows the data elements (columns) in each record (row) as they initially exist in the campaign database.

```
adoption_number (C and four digits for cats;
D and four digits for dogs)
last_name
first_name
prefix (Mr., Mrs. or Ms.)
pet_name
adoption_date
address
city
state
zip
```

Listing 8-1. Record Layout Before Changes

The Process

We will keep this simple, but the general idea should come through. The data in each row must match the document design. For each data location in the design layout, there has to be a corresponding column in the database table. (*More about this in Chapter 9.*) Let's begin supplementing data with new fields – there will be eight:

1. **tracking_code** – This field will be used to track responses. The content will be created by concatenating the adoption number, the zip code, and the start date for round one of the campaign (example: "D1234 14532 090615"). On the document it will be converted to a standard barcode by applying a font.
2. **mailing_code** – To take advantage of mailing rates, this field will be loaded with the addressee's zip code. It will be converted on the document to a mail barcode by applying another font.
3. **salutation** – The salutation will be created by concatenating "Dear " with prefix, last name, followed by a comma (example: "Dear Mrs. Smith,"). If the prefix is empty, the salutation will use the first name (example: "Dear Georgia,").
4. **addressee** – We will build the addressee by concatenating the prefix (if not empty), the first name and last name (example: "Ms. Ann Gorah").
5. **location** – Just as we did in the preceding field, we'll combine the city, state, and zip into one field to simplify processing (example: "Anyville, OH 45678").
6. **pet_image** – This field will contain the storage address of the digital photo. Its file name is adoption number, pet's name, and file type. We'll prepare for XML export by combining "href = file://photos/" with the file name (example: "href = file://photos/C3456Blossom.jpg").
7. **pet_message** – If the first character of the adoption number is a "D", this field will have a dog-related message. If the first character of the adoption number is a "C", this field will have a cat-related message.
8. **adoption_message** – If the adoption date is less than one year, the message will be "Thank you for recently adopting " and the pet's name. If the adoption date is older than a year, the message will be "Would " plus the pet's name, plus " like a playmate?" (example: "Would Scruffy like a playmate?").

Suppose there are a thousand records. We've just added eight new fields per records. That means eight thousand fields need content. While it is relatively easy to add columns to a table, filling them with appropriate content might be more involved. We certainly don't want to do it by hand, so, how do we automate?

For example, if we are using FileMaker Pro, we can set up scripts to do the work. The scripts can be built by pointing and clicking from a menu to create a logical sequence of commands. The script can be named and saved for repeated use.

If we are using Microsoft Access, we can create update queries by pointing and clicking. The queries can be saved for reuse. Both are simple, but the queries or scripts need thorough testing and validating of test results with a subset of the data before running against the entire dataset.

How many kinds of scripts must we build to provide the additional data for the dogs and cats campaign? Well, at this point the problem is primarily populating empty data columns in our table. The watchword is "simplicity". We can do all the necessary data manipulation to populate the eight new fields (listed above) with just four basic user-defined functions:

1. Inserting new text or values in every row of a particular column
2. Selectively inserting new text or values in one column based on the value of another column in the same row
3. Copying the content of each column and pasting it into another column in the same row
4. Assembling multiple column values into a single string and pasting it into another column in the same row

The scripts we will use are all variations on a theme – nothing really new. In fact, to emphasize this point, we will use functions originally defined for another application (see Chapter 19). The logic is already there; all we need to do is copy each one into our dogs & cats adoption database and change table and field names, plus any text strings. Figure 8-6 demonstrates a text insert function. We just have to do is modify the layout, field, and insert statements to reflect the adoption database and the modifications we want to make. We don't have to change any processing logic.

```
Script Name:  Field Text Insert

  Go to Layout ["Address"]
  Go to Field [Address::image]
  Loop
    Insert Text [Select; Address::offer; "5% Discount on purchases over $25.00"]
    Go to Record/Request/Page [Next; Exit after last]
  End Loop
```

Figure 8-6. Script for Inserting Column Values

The function in Figure 8-7 is a little more involved, but basically the same as the previous one, except that it inserts the text string into the specified column selectively if the value of another column meets specified criteria. In the case of the adoption database, selection criteria could easily be "D" for dog and "C" for cat.

```
Script Name:  Field Image XML Setup

  Go to Layout ["Address"]
  Go to Field [Address::image]
  Loop
    If [Address::title = "Ms."]
      Insert Text [Select; Address::image; "href = file://Blossoms.jpg"]
    End If
    Go to Record/Request/Page [Next; Exit after last]
  End Loop
```

Figure 8-7. Script for Selective Value Insertion

By this time, our column processing function should look very familiar. So as not to disappoint, Figure 8-8 performs the same positioning and looping process. The functional difference is that it copies the value from one column and then pastes that same value into another column.

```
Script Name: Field Copy Paste

Go to Layout ["Address"]
Go to Field [Address::image]
Loop
   Copy [Select; Address::city]
   Paste [Select; Address::location]
   Go to Record/Request/Page [Next; Exit after last]
End Loop
```

Figure 8-8. Script to Copy and Paste Values

The last script example (Figure 8-9, next page) follows the same mundane processing pattern. The new piece of logic takes values from multiple columns and concatenates (combines) them into a single value – in this case, a text string that will ultimately be the basis for a barcode. That barcode could just as easily be the adoption tracking code for the adoption database.

```
Script Name: Field Text Assembly

Go to Layout ["Address"]
Go to Field [Address::image]
Loop
   Insert Calculated Result [Select; Address::code; Address::zip & "-" & Address::number & "200803"]
   Go to Record/Request/Page [Next; Exit after last]
End Loop
```

Figure 8-9. Script to Combine Multiple Values and Then Insert the Result

Now that we have examined improving and setting up the data so that it contains all the information we will need for document design integration, it's time to begin the last steps in moving the data out of the desktop DBMS to create a data stream that our design software can use.

Content Stream Export

The reason for exporting the database content to XML is that the XML content stream facilitates automation of variable content document generation using Adobe InDesign. XML also assists in data quality checking, document design standardization, and design template creation.

The primary reason this discussion centers on FileMaker Pro is that it provides very clean XML exports requiring minimal cleanup prior to Adobe InDesign import. Cleanup in this context is not getting rid of "dirty" or incorrect data. It is streamlining the XML content stream. We will discuss this further in the next chapter.

Tag Naming

As a first step, regardless of what DBMS software is in use, review the table being exported in "define" or "design" mode. All table headings (field names) are going to become XML tag names. To make life simpler in anticipation of the XML conversion, change all characters in all the table headings to lower case. XML is case-sensitive; making all the XML tags lower case reduces the chance of confusion or error when later working with the XML content stream.

Configuring

In FileMaker Pro from the "File" menu, select "Export Records". In the initial panel, all we have to do is give the export file a name, provide a location for the export, and select "XML" as the type of export. Save and continue.

The next panel asks what XML grammar we want to use. Common sense would tell us to use "FMPXMLRESULT". After all, it says that the result is XML, right?. If we used this selection, we would have substantially more streamlining work to do, so common sense and intuitive choices don't always work.

Rather, choose "FMPDSORESULT" (think of DSO as "data set order") and make life simpler for these reasons: superfluous XML is minimized, individual rows are tagged, and the XML tags around content use the column (or field) name. An XSL stylesheet is not necessary, so continue.

Selecting

The last panel provides the capability to select what we want to export. Select the "Current Table" – we should have the table we have prepared as the active table prior to starting the export. We don't need to export everything, only those fields that will be used to populate the documents. Select the desired fields by highlighting them and either double-clicking each choice or using the move button.

Organizing

Continuing with preparations, the individual record field order within each content stream record must coincide with the order of the data as it will appear in the design layout. The order that the XML content stream loads into the document design will be top-to-bottom and left-to-right – meaning generally diagonally from the upper left corner to the lower right corner of the layout.

Reordering fields in File Maker Pro can be accomplished as part of the export preparations. Selected fields can be reordered up or down by clicking and dragging. When setup is completed – all fields required by the document design have been selected and placed in "design" order – select the "Export" button, and go on to Chapter 9 to prepare the XML.

If using another DBMS that only provides query-based data selection, the order of the individual data elements within the query determines content order in the resulting data extract, and Chapter 9 will take on new significance.

Supplemental Reading

There are newer editions available for this book. There is also an increasing number of related books on the FileMaker Pro DBMS. Also, there is an increasingly large number of books on Microsoft Access.

Coffey, G. & Prosser, S. (2005). *FileMaker Pro: the missing manual.* Sebastopol, CA: O'Reilly Media, Inc.

Data warehousing, data mining, and knowledge management are certainly applicable to the planning and generation of variable content documents. The following three books are a good starting point for getting a more complete understanding of these topics. Although older, they are very informative because they deal with the logic and concepts rather than technology specifics. Therefore, it is not necessary to reference the most leading edge database mechanics.

Mattison, R. (1999). *Web warehousing and knowledge management.* NY: McGraw Hill.

Rud, O. (2001). *Data mining cookbook: modeling data for marketing, risk, and customer relationship management.* NY: John Wiley & Sons, Inc.

Westphal, C. & Blaxton, T. (1998). *Data mining solutions: methods and tools for solving real–world problems.* NY: John Wiley & Sons, Inc.

9. Techniques – Managing XML Content Streams

Discussion Points:

- ✓ Next Step
- ✓ About The XML Content Stream
- ✓ Using A Text Editor
- ✓ Using XSLT
- ✓ Using Perl Programming
- ✓ Recommended Reading

Next Step

At this point, we have acquired the data for a variable document, prepared the data, and transformed it into XML. We've done a lot, but we still don't have anything that we can easily and successfully use to help populate our Adobe InDesign document layout. Figure 9-1 describes the steps involved in preparing the XML content stream. Starting with the XML export, we'll discuss optional approaches to structure and refine it.

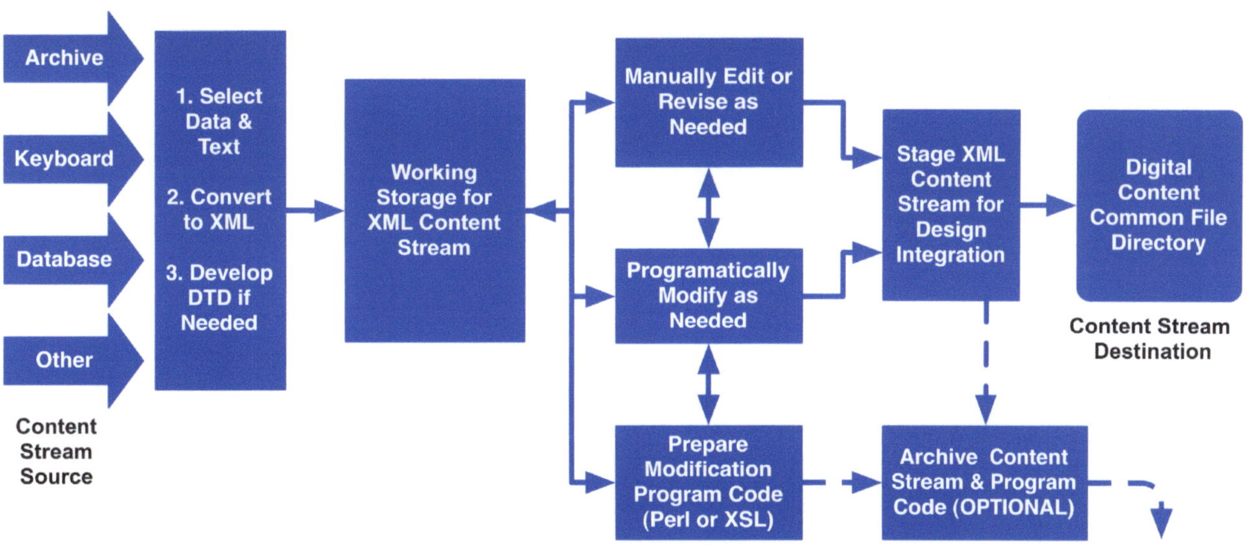

Figure 9-1. XML Content Stream Preparation Workflow

First we need to review the data we'll use. Listing 9-1 (top of the next page) is a small subset of the "Friends of Dogs and Cats" database. It has three rows (records), and each row has six columns (fields). This will work just fine. Let's export this data to start preparing our XML content stream.

number	name	breed	gender	photo	weight
1	Slinky	German Shepherd	Male	file://slinky.jpg	115
2	Nora	Doberman Pinscher	Female	file://nora.jpg	90
3	Crunch	Rottweiler	Male	file://crunch.jpg	125

Listing 9-1. The Dog Database

About the XML Content Stream

If the database export had been based on the FMPXMLRESULT grammar, we would have gotten an XML file containing 76 elements in a single string of characters. Listing 9-2 shows a portion of this file viewed in an XML editor.

```
<?xml version="1.0" encoding="UTF-8" ?>
<FMPXMLRESULT xmlns="http://www.filemaker.com/fmpxmlresult">
<ERRORCODE>0</ERRORCODE>
<PRODUCT BUILD="11-30-2007" NAME="FileMaker Pro" VERSION="8.5v2"/>
<DATABASE DATEFORMAT="M/d/yyyy" LAYOUT="" NAME="DogDB.fp7" RECORDS="3" TIMEFORMAT="h:mm:ss
a"/>
<METADATA>
    <FIELD EMPTYOK="YES" MAXREPEAT="1" NAME="number" TYPE="NUMBER"/>
    <FIELD EMPTYOK="YES" MAXREPEAT="1" NAME="name" TYPE="TEXT"/>
    <FIELD EMPTYOK="YES" MAXREPEAT="1" NAME="breed" TYPE="TEXT"/>
    <FIELD EMPTYOK="YES" MAXREPEAT="1" NAME="gender" TYPE="TEXT"/>
    <FIELD EMPTYOK="YES" MAXREPEAT="1" NAME="photo" TYPE="TEXT"/>
    <FIELD EMPTYOK="YES" MAXREPEAT="1" NAME="weight" TYPE="NUMBER"/>
</METADATA>
<RESULTSET FOUND="3">
<ROW MODID="2" RECORDID="1">
        <COL> <DATA>1</DATA> </COL>
        <COL> <DATA>Slinky</DATA> </COL>
        <COL> <DATA>German Shepherd</DATA> </COL>
        <COL> <DATA>Male</DATA> </COL>
        <COL> <DATA>file://slinky.jpg</DATA> </COL>
        <COL> <DATA>115</DATA> </COL>
</ROW>
```

Listing 9-2. The Top plus One Record of an FMPXMLRESULT Export

There is really nothing workable in Listing 9-2. Almost all of the top (red highlights) has to be discarded, all the records will be like the example in yellow highlights; it will require significant rework to get the XML we need. All we want is an XML file that presents the three database records with some similarity to how they look in the DBMS – and that includes column names converted to XML tags.

The export using the FMPDSORESULT grammar generates a single string of characters representing only 30 XML elements. There is still some cleanup to do, but the result is something considerably more usable. Listing 9-3 on the following page shows the complete export (all three records) as viewed in the XML Editor.

Other than deleting four XML elements (red highlights) and cleaning up the root and row tags (yellow highlights), the content stream is almost ready to go. After this, the only cleanup left is to convert the content of the photo element (green highlights) into an "href" attribute in an empty XML element. A text editor could easily accomplish this final preparation step.

```
<?xml version="1.0" encoding="UTF-8" ?>
  <!-- This grammar has been deprecated - use FMPXMLRESULT instead -->
  <FMPDSORESULT xmlns="http://www.filemaker.com/fmpdsoresult">
    <ERRORCODE>0</ERRORCODE>
    <DATABASE>TestDB.fp7</DATABASE>
    <LAYOUT/>
    <ROW MODID="2" RECORDID="1">
      <number>1</number>
      <name>Slinky</name>
      <breed>German Shepherd</breed>
      <gender>Male</gender>
      <photo>file://slinky.jpg</photo>
      <weight>115</weight>
    </ROW>
    <ROW MODID="0" RECORDID="2">
      <number>2</number>
      <name>Nora</name>
      <breed>Doberman Pinscher</breed>
      <gender>Female</gender>
      <photo>file://nora.jpg</photo>
      <weight>90</weight>
    </ROW>
    <ROW MODID="0" RECORDID="3">
      <number>3</number>
      <name>Crunch</name>
      <breed>Rottweiler</breed>
      <gender>Male</gender>
      <photo>file://crunch.jpg</photo>
      <weight>125</weight>
    </ROW>
  </FMPDSORESULT>
```

Listing 9-3. FMPDSORESULT Export Results (all three records)

When all cleanup is completed, all the superfluous XML elements are gone. All the unneeded XML element attributes are gone. All the XML tags are in lower case (for our convenience), and all extra white space (primarily indenting) has been removed. The references to the external digital images have been converted to href element attributes as the InDesign processor expects.

The final objective, as shown by Listing 9-4 (next page), will be a completely clean XML content stream ready for import into Adobe InDesign.

How did we do this? What works? What doesn't? How much work is entailed? We'll look at three approaches: text editor, XSLT (extensible stylesheet language transform), and Perl programming. Each approach will result in Listing 9-4.

Using a Text Editor

A text editor is quite adequate for one-time deletions such as the lines highlighted in red in Listing 9-3. Modifying or renaming values that appear in multiple places throughout the content stream can also be done using the text editor "Search & Replace" capability. This includes changing cases and eliminating indents (white space added to enhance readability).

A text editor can also be used to convert tagged content into an XML attribute. This is the type change required to access external digital images during the content stream import into the design layout. In our "Dog" XML content stream, there are digital image addresses we want to use. Follow along as we prepare all the XML "photo" elements like this sample line extracted from Listing 9-3:

```
<photo>file://nora.jpg</photo>
```

```
<?xml version="1.0" encoding="UTF-8" ?>
<dataroot>
<row>
<number>1</number>
<name>Slinky</name>
<breed>German Shepherd</breed>
<gender>Male</gender>
<photo href="file://slinky.jpg"/>
<weight>115</weight>
</row>
<row>
<number>2</number>
<name>Nora</name>
<breed>Doberman Pinscher</breed>
<gender>Female</gender>
<photo href="file://nora.jpg"/>
<weight>90</weight>
</row>
<row>
<number>3</number>
<name>Crunch</name>
<breed>Rottweiler</breed>
<gender>Male</gender>
<photo href="file://crunch.jpg"/>
<weight>125</weight>
</row>
</dataroot>
```

Listing 9-4. Clean XML Content Stream

1. Using search and replace, we select all the leading photo tags

    <photo>

2. And replace each one with

    <photo href="

3. Then we select all the photo end tags

    </photo>

4. and replace each with

    "/>

5. Making the results

    <photo href="file://nora.jpg"/>

This is exactly what we need in order to process the digital images. Some preplanning during data preparation while still in the DBMS is needed to facilitate this – primarily adhering to a standard pattern. Scripting using DBMS capabilities will help.

What You Need to Know

Both "Mac" and "Windows" operating systems come with their own text editors. Either can do most of the work. For more text editing capabilities, a programmer's editor (BBEdit, TextWrangler, or Komodo, for instance) or an XML editor such as SyncRO Soft's <oXygen/> will be able to handle more complex search and replace processing that could require pattern matching and the use of "wild cards" or place holders (for instance, an asterisk or question mark).

Consider this problem. An XML tag has two attributes. Each attribute has a different value. There are 500 such tags, each designating a different row We need to drastically simplify all *the row tags so that every one is just <row>*. We may well need more matching capabilities than what most text editors can handle just to make these simple row changes.

Another thing you need to know is that changes may require multiple passes through the data. Setting up the digital image required two. If there were three levels of indents in the XML, three passes through the data may be needed to eliminate the leading white space. Doing a search and replace has its risks. The more passes through the data, the more chance of a misstep – especially if passes need to be executed in a specific sequence.

Given the possibility for errors, one or two trials replacing one occurrence at a time is a prudent step before clicking the "Replace All" button. Of course you could just close the file without saving and then try again. The question is, "Did you save the file each time you made a change so you don't have to start at square one again?"

Using XSLT

From the discussion on text editors, it's apparent they do not provide a complete solution to XML content stream cleanup. So why not use capabilities built into the XML technology – XSLT (extensible stylesheet language transforms)? XSLT uses a construct called the Document Object Model (DOM for short) to do the processing. DOM reads the entire content stream into memory according to its XML hierarchy. It then repetitively traverses the hierarchy performing instructions from an extensible stylesheet language (XSL) document.

Figure 9-2 describes the basic XSLT process. The XSL instruction sheet is read into the XSLT processor. The XSLT processor then reads into memory the XML source file and performs the XSL instructions one at a time on the entire XML file residing in memory. When all XSL instructions have been executed, the XSLT processor generates a new revised XML file. The original source file is not modified.

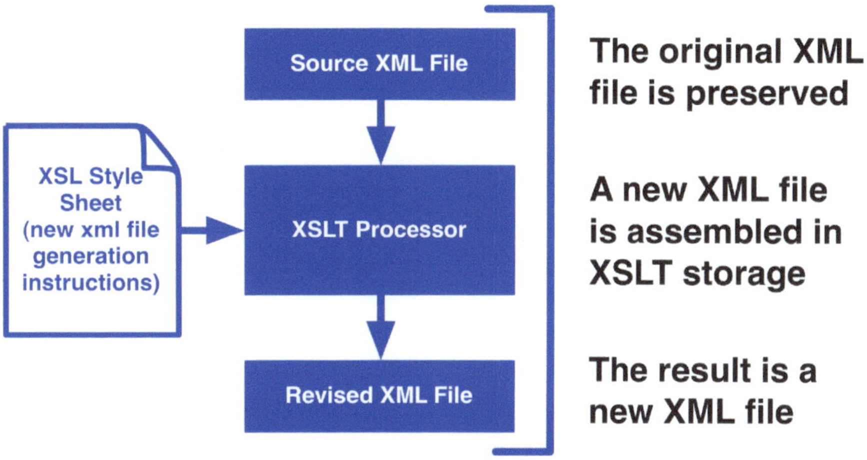

Figure 9-2. XSLT Process Overview

Figure 9-3 presents a conceptualized DOM architecture as stored by the XSLT processor, and basic processes it executes according to XSL instructions to modify the XML content stream.

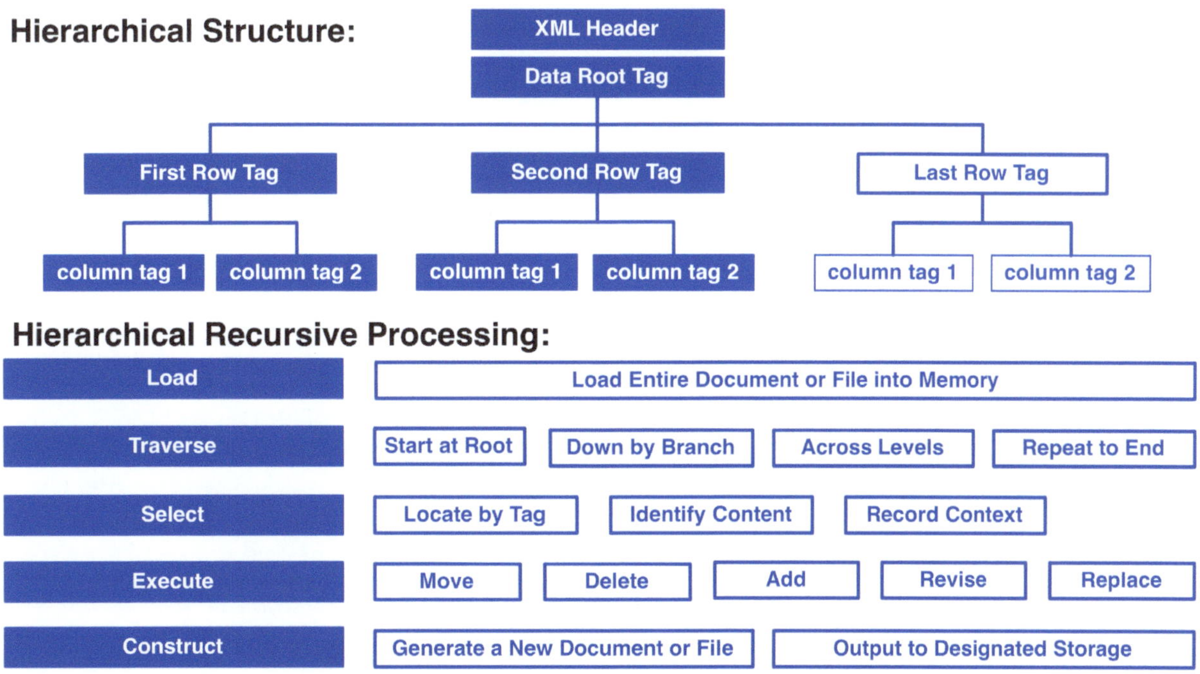

Figure 9-3. XML DOM Processing Model

Let's continue starting with the raw XML that was generated by FileMaker Pro. Reference Listing 9-1 and Listing 9-3 if you need to refresh your memory. Our goal is to massage the XML file into the same exact structure presented earlier in Listing 9-4.

The first need is to get the raw content stream ready to process. It is best to use an XML editor (SyncRO Soft's <oXygen/> is one that is cross-platform capable – both PC and Mac). The raw XML content stream will come into the editor as a single uninterrupted string – no line breaks. We, begin by asking "<oXygen/>" to apply formatting, line breaks, and indents. The XML will be easier to understand and more workable.

Then we'll use our text editor to delete the attribute (highlighted in yellow) from the root element shown below:

```
<FMPDSORESULT xmlns="http://www.filemaker.com/fmpdsoresult">
```

Resulting in

```
<FMPDSORESULT>
```

Now, we will use XSLT for the rest of our preparations. We'll use two XSL stylesheets to address the our basic processing requirements: the first to simplify and structure the content stream; the second to convert image addresses into reference attributes that Adobe InDesign can use to pull digital images into the document layout.

The first stylesheet (Listing 9-5) will recursively read the raw XML content stream, perform a set of cleanup instructions, and then create an intermediate XML file.

The easiest way to understand the cleanup stylesheet (Listing 9-5) is to dissect it. The first line is the XML declaration. This is followed by the XSL root and namespace declaration, output instructions, and an instruction to strip away superfluous white space.

Continuing to work with Listing 9-5, we see three processing instruction sets, each preceded by an explanatory comment (yellow highlight). The first instruction set removes unneeded XML elements.

The second instruction set finds all <ROW MODID="2" RECORDID="1"> elements and replaces them with a simple set of "<row> ... </row>" tags without attributes.

The last instruction set replaces the original document root element, "FMPDSORESULT", with a more understandable tag set, "<dataroot> ... </dataroot>". And the last line is the closing tag for the stylesheet, XSL root.

```
<?xml version="1.0" encoding="UTF-8"?>
<xsl:stylesheet xmlns:xsl="http://www.w3.org/1999/XSL/Transform"
version="1.0">
<xsl:output method="xml" version="1.0" encoding="UTF-8" indent="yes"-
/>
<xsl:strip-space elements="*"/>

<!-- Remove superfluous xml elements -->
<xsl:template match="ERRORCODE"/>
<xsl:template match="DATABASE"/>
<xsl:template match="LAYOUT"/>

<!-- Copy ROW Children into new row parents -->
<xsl:template match="ROW">
   <row>
      <xsl:copy-of select="*|node()"/>
   </row>
</xsl:template>

<!-- Replace FMPDSORESULT root with a new root -->
<xsl:template match="/">
   <dataroot>
      <xsl:apply-templates/>
   </dataroot>
</xsl:template>

</xsl:stylesheet>
```

Listing 9-5. The Raw XML "Cleanup" Stylesheet

The second stylesheet (Listing 9-6) imports the intermediate XML file generated from the first stylesheet. Then it creates image reference attributes, resolves XML element ordering, and creates a near-complete XML content stream. This stylesheet has the same structure as the previous one. We will continue highlighting instruction set comments in yellow.

The first instruction set adds a new "photo" element containing a file reference attribute for the digital image, and a new trailing "weight" element. This means that we have two "photo" and two "weight" elements – the new pair and the original pair.

The next instruction set removes the old photo and weight elements.

The last instruction set, as its comment says, copies everything else. With the XML editor's search and replace capabilities, eliminate any leading indents that might remain. The completed XML content stream is ready. It should look like Listing 9-4 earlier in the chapter.

```
<?xml version="1.0" encoding="UTF-8"?>
<xsl:stylesheet xmlns:xsl="http://www.w3.org/1999/XSL/Transform"
version="1.0">
<xsl:output method="xml" version="1.0" encoding="UTF-8"
indent="yes"/>
<xsl:strip-space elements="*"/>

<!-- Add new image element with reference attribute -->
<!-- Add new weight element following image element -->
<xsl:template match="row">
    <row>
      <xsl:apply-templates/>
       <photo href="{photo}"/>
       <weight><xsl:value-of select="weight"/></weight>
    </row>
</xsl:template>

<!-- Remove old duplicate elements -->
<xsl:template match="photo"/>
<xsl:template match="weight"/>

<!-- Copy everything else -->
<xsl:template match="@*|node()">
    <xsl:copy>
       <xsl:apply-templates select="@*|node()"/>
    </xsl:copy>
</xsl:template>

</xsl:stylesheet>
```

Listing 9-6. The Image Element "Conversion" Stylesheet

What You Need to Know

Although XSLT is a powerful and complex technology with capabilities well beyond what we are using, a complete understanding of the technology is not necessary to be productive in preparing the XML content stream. This implies that you may encounter some frustration from trial-and-error attempts at achieving your desired results. However, having the right tools to do the job is helpful. While "field expedients" may work, they may also introduce added frustrations. An XML editor is an important tool. For our purposes, it is easy to learn. After all, XML is text. An XML editor is essentially a text editor with all kinds of powerful XML-related capabilities.

Standardization is important. While data content and data elements names are going to be different, the XML structure is generally the same or similar between jobs. This means that the two stylesheets for the 3-record dog content stream will work with a 2,000-record customer content stream with only minuscule accommodation – generally different tag names and additional digital image references. Changing tag names in the stylesheet is simple. Dealing with added digital image references is only a matter of copying the instruction lines for the "dog photo" reference.

Using Perl Programming

This brings us to our third processing option – using a Perl program. Perl is the acronym for "Practical Extraction and Reporting Language". We're not going to do any reporting. Instead we'll use Perl's powerful text manipulation capabilities included in its basic functionality.

By the way, the objective of this discussion is to demonstrate what Perl can do, not to teach teach you how to program with Perl.

What we are going to do is process the XML content stream, newly exported from the desktop DBMS, as if it were a sequential file. But wait! Stop! This doesn't make sense! In the previous chapter (Chapter 8), a lot of effort went into explaining the problems with sequential file processing, and how using a visual matrix and DBMS was better. Now the talk returns to sequential file processing. It all started with a sequential file that was transformed into a database, exported into XML, only to be treated as if it were a sequential file. Talk about going around the barn!

Here's the difference. Remember, we started with raw data where records and fields were primarily identified by character counts or some sort of delimiters. Now we are dealing with refined data that has been reorganized into a sequential structure where each data field is identified by a set of XML tags. The XML tags provide a uniform "roadmap" through the entire data set from start to finish. We'll read the file sequentially, stopping at selected tags along the way to do cleanup.

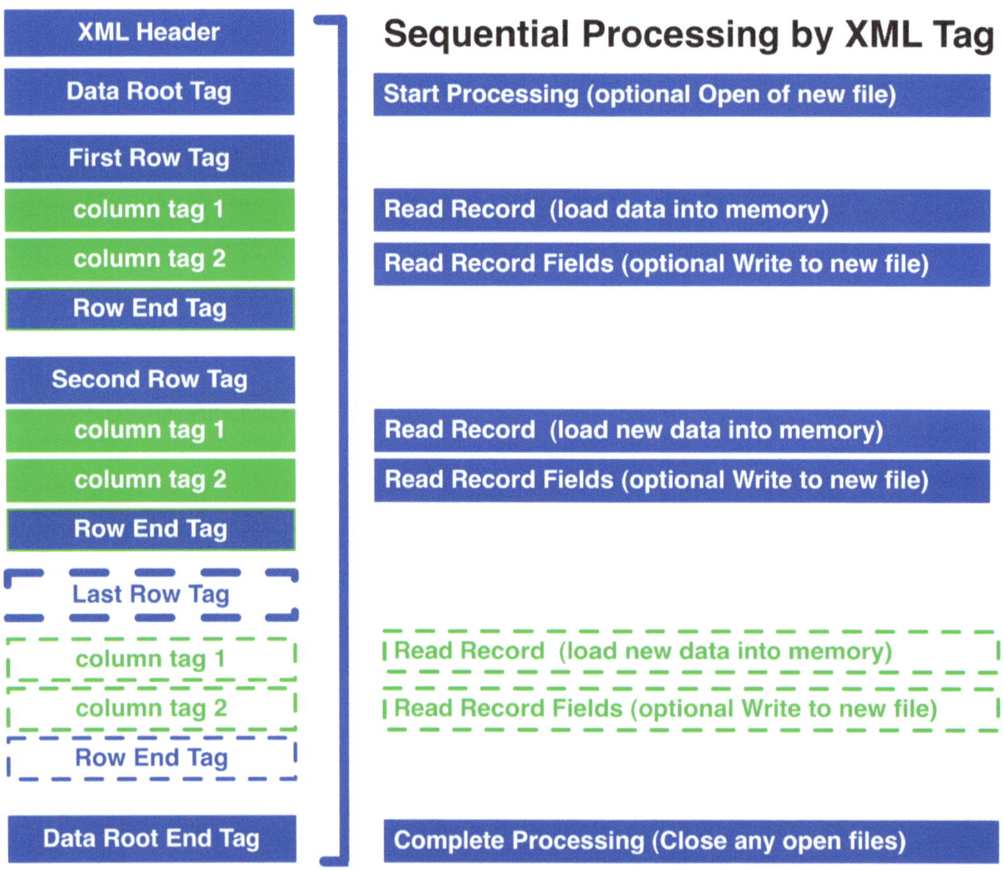

Figure 9-4. XML Sequential Processing Model

Figure 9-4 (above) probably looks familiar. It is! It is essentially the same sequential file model we discussed before, except that it is stringently organized with XML tags according to XML conventions. Starting with the first XML element we read each tag or tag-content combination and decide what to do with it.

Figure 9-5 demonstrates how this process works. It is a relatively low-level picture of what happens. We're looking at one tag set (or element) at a time. Read Figure 9-5 from top left across, and then advance to the next line, starting again on the left.

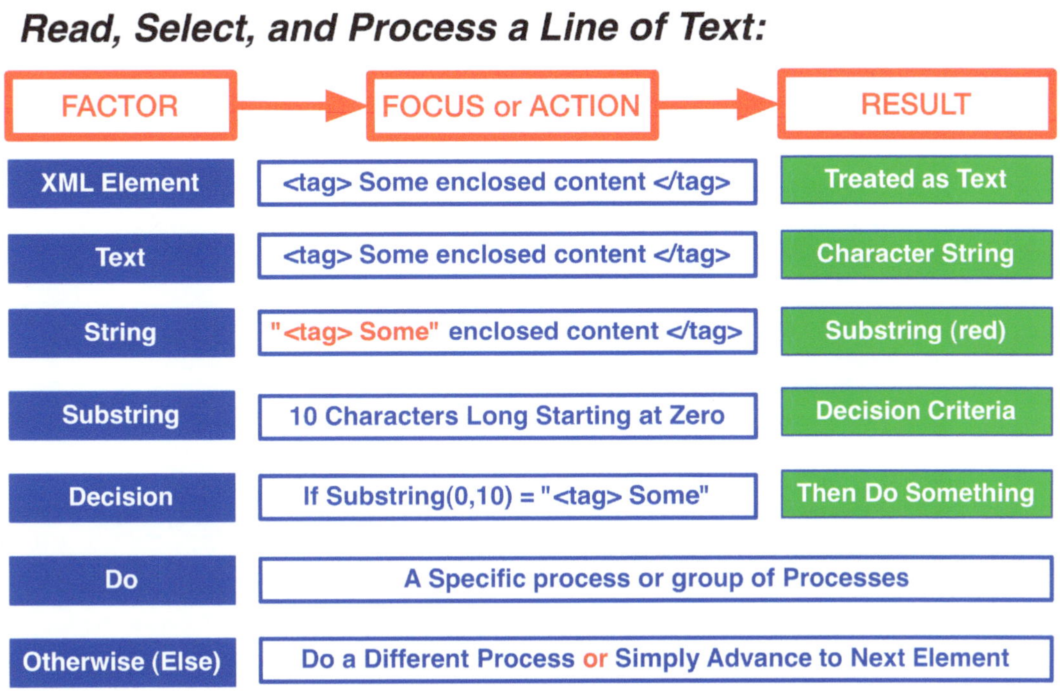

Figure 9-5. Sequential XML Processing Logic

By adding a little altitude, we lose some detail but gain better perspective (the proverbial forest and trees). We find that we have been looking at one step in a processing loop demonstrated by Figure 9-6 below. Our knowledge of how we will conceptually process the XML content stream helps us describe the process to the computer so it can perform the actual work.

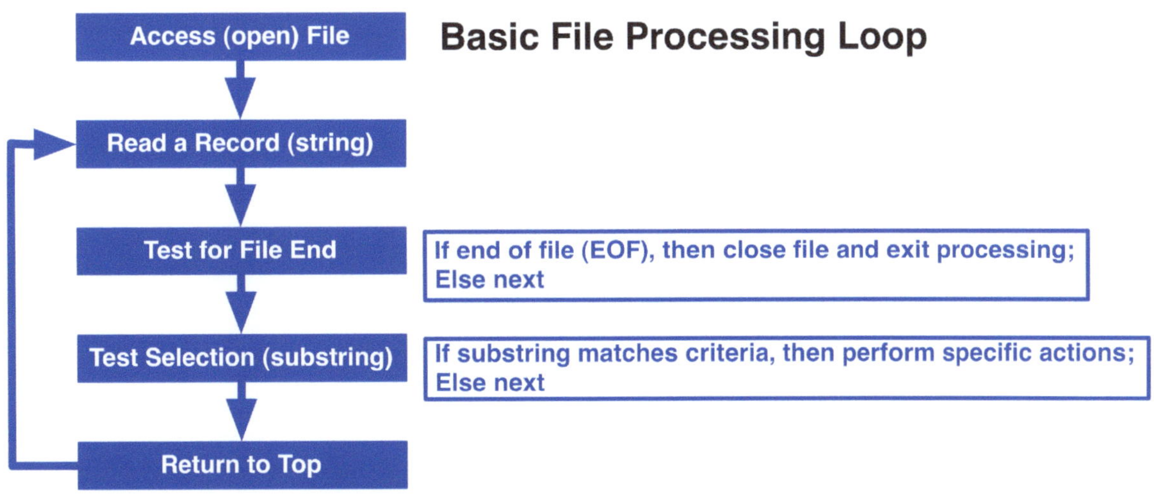

Figure 9-6. Sequential XML Processing Flow

A single program is capable of doing all the cleanup we previously discussed using a text editor or XSLT, including element-to-attribute conversion. We'll package related functionality into standard blocks called "subroutines". (*This approach has some conceptual similarities to how our XSLT was organized.*)

Looking at actual program code will help us better understand the programming concepts presented here, while addressing our primary objective – preparing the XML content stream. Let's dissect this program. Just follow along closely; you'll be fine without knowing any programming nitty-gritty. Figure 9-7 provides a basic model for all the Perl programs we will examine; they all follow this same pattern.

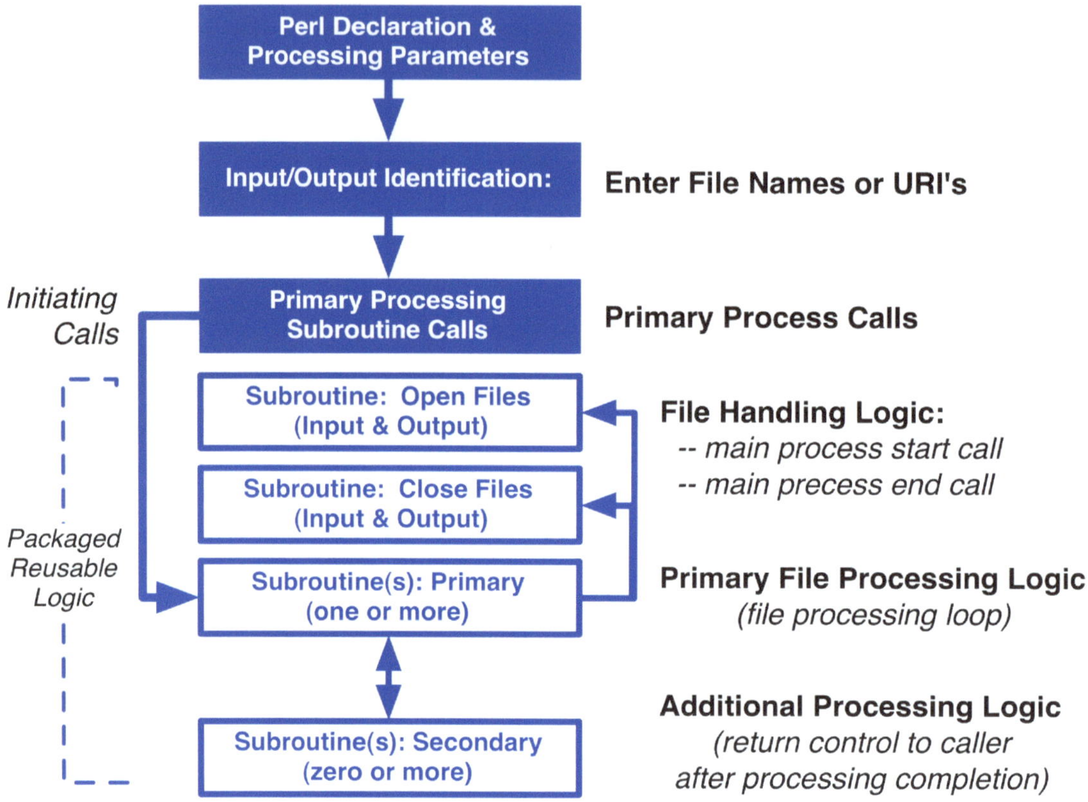

Figure 9-7. Basic Program Architecture

To start, the first line of every executable Perl program (at least for UNIX, Linux, and Mac OS X systems) starts with #! (called a "shebang" and shown below). The line helps the operating system find the Perl interpreter and associated components. PC's rely on operating system preference settings and file extension associations to accomplish the same thing – the "shebang" isn't needed.

```
#!/usr/bin/perl
```

Next are two processing parameters. One requests warnings for potential issues. The other defines how the program interpreter should handle subroutines.

```
use warnings;
use strict "subs";
```

The next lines set variables used by the program. The first two identify the name and location of input and output files. The third identifies a temporary output file. The last two are constants for element-to-attribute conversion.

```
$inputfl = "rawXMLin.txt";
$outputfl = "cleanXMLout.xml";
$transfl = "deleteME.xml";
$href = "href=\"";
$endtag = "\"/>";
```

Two statements initiate primary processing. The first executes a subroutine to ready XML for sequential processing by inserting line breaks between elements. The second executes the cleanup.

```
&initializeXML;
&cleanseXML;
```

The next set of instructions is a subroutine to open the input file for reading, and open (or establish) the output file for printing. "Printing" simply means writing to a destination – in this case, a text file on disc.

```
sub openFiles {
    $infl = $_[0];
    $outfl = $_[1];
    open(INFILE, "<$infl") || die "Can't find
    or open $infl: $!";
    open(OUTFILE, ">$outfl") || die "Can't
    open $outfl: $!";
}
```

At process completion, any open files need to be closed. This is like reading a book and taking notes in a journal. You open the book and journal to start reading and note taking. When finished, you close both. This subroutine is called upon as the last step in file processing to close both the input file and output files. It releases the files and makes them available for later reuse.

```
sub closeFiles {
    close(INFILE);
    close(OUTFILE);
}
```

The XML content stream exported from the DBMS is one contiguous string of text. We need to break that string so that each XML element is a single line.

The initialize subroutine facilitates sequential file processing. It simply places a new line code "\n" at the juncture of every XML end and start angle bracket. For example, it changes "<row><breed>" to "<row>\n<breed>".

```
sub initializeXML {
    &openFiles($inputfl, $transfl);
    while ($import = <INFILE>) {
        chomp($import);
        $import =~ s/></>\n</g;
```

```
#!/usr/bin/perl

use warnings;
use strict "subs";

$inputfl = "rawXMLin.txt";
$outputfl = "cleanXMLout.xml";
$transfl = "deleteME.xml";
$href = "href=\"";
$endtag = "\"/>";

&initializeXML;
&cleanseXML;

sub openFiles {
    $infl = $_[0];
    $outfl = $_[1];
    open(INFILE, "<$infl") || die "Can't find or open $infl: $!";
    open(OUTFILE, ">$outfl") || die "Can't open $outfl: $!";
}

sub closeFiles {
    close(INFILE);
    close(OUTFILE);
}

sub initializeXML {
    &openFiles($inputfl, $transfl);
    while ($import = <INFILE>) {
        chomp($import);
        $import =~ s/></>\n</g;
        $export = $import . "\n";
        print OUTFILE $export;
    }
    &closeFiles;
}
```

Listing 9-7. Perl Program for XML Content Stream Cleanup (continued on next page)

```
            $export = $import . "\n";
            print OUTFILE $export;
        }
        &closeFiles;
}
```

Now we are ready to do the remaining 95% of the XML content stream cleanup. Most cleanup is really simplifying the XML by getting rid of unneeded elements or attributes. The cleansing subroutine performs this series of processes based on each type of XML element encountered.

Before stepping through this subroutine, it may be helpful to refer back to Listing 9-3, "FMPDSORESULT Export Results". Keep this listing in mind as we continue.

We'll use the continuation of Listing 9-7 just to the right. The subroutine we're interested in is called, "cleanseXML" – the first line of code (highlighted). Here we go, step by step:

```
sub cleanseXML {
        &openFiles($transfl, $outputfl);
        while ($import = <INFILE>) {
                chomp($import);
                if (substr($import,0,4) eq "<ROW"){$import = "<row>";}
                elsif (substr($import,0,6) eq "</ROW>") {$import = "</row>";}
                elsif (substr($import,0,4) eq "<!--") {next;}
                elsif (substr($import,0,13) eq "<FMPDSORESULT")
                {$import = "<dataroot>";}
                elsif (substr($import,0,11) eq "<ERRORCODE>") {next;}
                elsif (substr($import,0,10) eq "<DATABASE>") {next;}
                elsif (substr($import,0,8) eq "<LAYOUT>") {next;}
                elsif (substr($import,0,9) eq "</LAYOUT>") {next;}
                elsif (substr($import,0,15) eq "</FMPDSORESULT>")
                {$import = "</dataroot>";}
                elsif (substr($import,0,7) eq "<photo>")
                {$import = hrefElement($import);}
                $export = $import . "\n";
                print OUTFILE $export;
        }
        &closeFiles;
}

sub hrefElement {
        my $element = $_[0];
        $lentxt = length($element);
        $boundary1 = index($element,">");
        $tag = substr($element,0,$boundary1);
        $offset = $boundary1 + 1;
        $boundary2 = ($lentxt - rindex($element,"<"));
        $sublen = $lentxt - $offset - $boundary2;
        $content = substr($element, $offset, $sublen);
        $newelement = ($tag . " " . $href . $content . $endtag);
        return $newelement;
}
```

Listing 9-7 Perl Program (continuation)

1. Call the "open" subroutine to commence file processing
2. Enter the processing loop and read an XML line, eliminating trailing blanks
3. If this is a "ROW" element with attribute, replace it with a "row" element without attribute
4. If it is an end "ROW" element, replace it with an end "row" element
5. If it is a comment, skip over it (meaning, don't write it to output)
6. If it is the starting "FMPDSORESULT" tag, replace it with "dataroot" (We will do the same with the corresponding end tag when we encounter it.)
7. If we encounter "ERRORCODE", "DATABASE", or "LAYOUT" tags, we will skip them (not write to output)
8. If we encounter a "photo" tag, we rebuild it using our "hrefElement" subroutine, explained last
9. As we loop through input, we either write the the XML line as is, write the replacement, or don't write anything; then we read the next XML line
10. Upon reading all XML lines, we call the "close" subroutine to terminate file processing

The last subroutine in the program is called, "hrefElement". It reconstructs the "photo" element, shifting its content to the value of a new "href" attribute in an empty "photo" element. On input, the photo element looks like this:

```
<photo>file://nora.jpg</photo>
```

This process evaluates the element and then deconstructs it. The subroutine takes the separated pieces, uses some, and replaces others to reconstruct a well formed XML element with attribute. Here is the result:

```
<photo href="file://nora.jpg"/>
```

The program code to accomplish this is contained in the subroutine below. (It's the last subroutine in the continuation of Listing 9-7 on the prior page.) The majority of this logic is devoted to identifying, modifying, or moving snippets of text to create the new XML element and attribute. To get a better understanding of this, it's helpful to compare Listing 9-3 with Listing 9-4.

```
sub hrefElement {
    my $element = $_[0];
    $lentxt = length($element);
    $boundary1 = index($element,">");
    $tag = substr($element,0,$boundary1);
    $offset = $boundary1 + 1;
    $boundary2 = ($lentxt - rindex($element,"<"));
    $sublen = $lentxt - $offset - $boundary2;
    $content = substr($element, $offset, $sublen);
    $newelement = ($tag . " " . $href . $content . $endtag);
    return $newelement;
}
```

When all program execution successfully completes, there remains an extraneous file called "deleteME.xml". This temporary file could have been deleted as part of the file close subroutine. But as stated earlier, we may need this file to help diagnose cleanup problems that were encountered. As a precaution, we'll plan on deleting it manually if it isn't needed – no big deal.

What You Need to Know

The good news is that Perl is easy to learn and the demonstrated program logic is just a series of loops and "if" statements. More good news is that the demonstrated program logic is reusable simply by changing some (not all) of the XML element tag names and file names. And finally, you do not need to know everything about programming or Perl to use this powerful tool.

The bad news is that you will have to use the terminal and issue text-based commands to run the Perl programs. You will need some smarts about your operating system beyond simply pointing and clicking. (There are plenty of books to help with this.) And, you will also need to get over the small hurdle of understanding Perl "shorthand". The recommended reading will help.

Appendix B is for those who would like to know more about altering XML content streams using the Perl approach and also avoid rework entailed by going back to the database to make changes. All the programs (including the one in this chapter) have been fully tested. And by the way, Appendix B also includes a Perl program that will, with a little setup, convert delimited files directly into an XML content stream without the database step. This may simplify the workflow, but setting up the program for a particular delimited file will be a little more demanding than what we've seen in this chapter.

Recommended Reading

For those who want more understanding of XML and what the discussion covered, the following two books are a good starting point. Wagner and Mansfield will provide an easy introduction. Stanek will provide more in-depth technical detail.

Stanek, W. (2002). *XML pocket consultant.* Redmond, WA: Microsoft Press.

Wagner, R. & Mansfield, R. (2003). *XML all-in-one desk reference for dummies.* NY: Wiley Publishing, Inc.

To get more comfortable with Perl, Schwartz and company is a great place to start. They provide plenty of examples and exercises (with solutions if you get stuck). Do the exercises, even if you have to rely on the solutions – that means write (or copy) the programs and then run them on you computer so you can see the results. Quigley will provide more details and more examples.

Quigley, E. (2008). *Perl by example, fourth edition.* Boston, MA: Pearson Education, Inc. and Prentice Hall.

Schwartz, R., Phoenix, T., & Foy, B. (2005). *Learning Perl, fourth edition.* Sebastopol, CA: O'Reilly Media, Inc.

10. Document Design: Integration & Print Stream

Discussion Points:

- ✓ A Rough Sketch
- ✓ Geometry
- ✓ Page Masters
- ✓ Styles
- ✓ Print Stream
- ✓ Recommended Reading

A Rough Sketch

As they say in orienteering or land navigation, "take a back azimuth" – development has moved from data acquisition through cleanup, enhancement, XML conversion, and XML simplification. Document design is next.

Our workflow explanation up to now has been sequential – one step at a time. We will continue that way for the sake of clarity. However, reality is somewhat ad hoc. The paradox in VCP is that the selection of data is dependent on design requirements; design requirements are dependent on the availability of appropriate data. Document design and data-related activities are actually concurrent and iterative – not the customary sequential steps usually portrayed as workflows.

What does this have to do with "rough sketches"? Just this! The development process is not implicitly consecutive. More likely, work will be done in parallel steps that are probably iterative. This suggests a high propensity for continuing change. This also suggests that more than one developer may be participating. Consequently, a simple vehicle is needed to record changes and carry design information forward. A rough sketch will serve as our roadmap, decision record, and initial communication devise.

Figure 10-1 (at the top of the next page) is a simple document blueprint showing what goes where. It is not complicated or according to scale (although it could be to scale). Recorded measurements are accurate. It could be rendered with pencil and paper, but changes and distributions could more easily be done with a digital tool such as Microsoft Visio (for the PC) or Omni Group's Omnigraffle Pro (for the Mac).

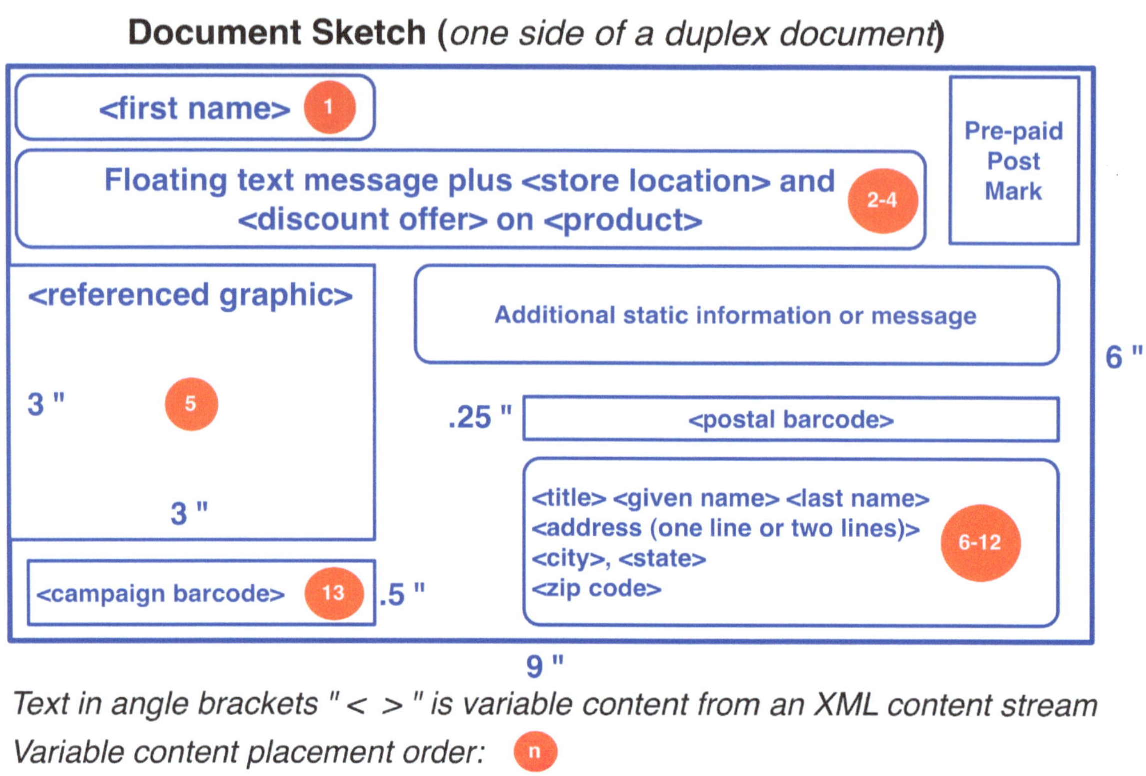

Figure 10-1. Design Geometry (Images, Data, and Text)

Geometry

Let's consider Figure 10-1 in more detail. It's more "engineering-like" and hardly artistic. It is a simple drawing with minimal distractions, and easy to modify as changes are identified. It's meant to answer basic questions about the document and its contents. The questions address the technicalities of automated document creation (process engineering).

Document Space

The objective here is to understand and validate efficient space utilization, and determine the overall relationship of all the design components.

- What are the outer dimensions and available space?
- What are the individual components?
- Where do the components go relative to each other?
- How much space does each component use?
- Do all the components comfortably fit in the space available?
- Is there remaining space to accommodate potential component size variances?
- What is the relative order of components from top to bottom and left to right?

Images

Next, we want to validate that the digital images fit into their designated spaces within the document. What we are doing is essentially standardizing the size of each type of image to its designated destination. Standardization simplifies automation.

- What are the dimensions of each digital image?
- Do the dimension of each image match the allocated space available?
- What images are static? (They will be the same on every generated document or page.)
- What images are variable? (They will vary from one page or document to the next.)
- Where are the variable images in relative order with all other variable components?

Content Stream

The purpose of these questions is to ensure that the content stream and the document sketch are coordinated. This is important because the document design software we're using (Adobe InDesign) processes each XML element once in the order it resides in the XML content stream.

- Is the XML content stream "well formed" and If a DTD is in use, is it "valid"?
- Do the XML "row" elements (child elements of the row tags) in the content stream reflect all the variable components (data and images) in the document sketch?
- Are the XML "row" elements in the same order as the variable elements portrayed in the document sketch (upper left to bottom right for each document page or side)?

Text

Other than XML data-related content, there are two other kinds of text: static text and floating text – what Maivald and Palmer (2008) call "cloned text". Static text remains the same from page to page or document to document. It might be page headers and footers, letterhead, a standard message, contact information, background images, or something else. The questions are few and simple:

- What is it, how big is it, and where is it placed?
- Does it vary by left or right page, document front or back, or is it always the same?
- Does it contain variable information related to the document – for instance, page numbers?

Floating or "cloned" text is static in the sense that, while associated with particular XML elements containing variable content, the floating text itself does not vary. Hence, it is "cloned" with each occurrence of a particular XML element. XML content loaded into a document, may likely be embedded within floating text. The text is "floating" because the space it occupies varies according to the character count of the embedded XML element. A long name, for instance, would create some expansion whereas a short name would cause contraction. Here are a few questions:

- What is the floating text and what is the associated XML element?
- What range of variance – "float" – can be expected for the content length of the associated XML element (most characters possible minus least characters possible)?
- What is the smallest and largest space required for the XML element and floating text combination?
- Is there enough unused space given the placement of all document components to accommodate the potential expansion and contraction?

Page Masters

Documents of every type and shape are structured and expressed as pages. Page masters provide structure; page spreads express content.

Page masters define the page layouts used in a document. Every page spread of a given layout originates from the page master for that layout. Page masters may define headers, footers, and page numbering. They may also define other static structures such as columns, master text frames, static text and image boxes. Page spreads, derived from page masters, are the page-by-page merger of structure and content.

That explanation is really a functional definition of page masters and page spreads. Now add content use with respect to page structure. Does the same content reappear on every page, every other page, or on particular page layouts? Is the content generally unique for each page?

As discussed in Chapter 3, "What Is a Variable Document", content that is unique to each page is "variable content", and content that repeats in some pattern throughout the document pages is "static content". For variable content publishing and document automation, static content is loaded onto page masters and variable content is loaded onto derived page spreads. When a page spread is derived from a specific page master, it takes everything residing on the master: structure, images, and text. Beside facilitating automation, this provides design continuity and time savings.

Styles

Styles not only determine how content looks on the page, they also determine how different content elements interact – for instance, how text wraps around an object or where pagination occurs. There are three types of styles:

1. Paragraph Styles – They are used to format blocks and strings of text.
2. Object Styles – They manage the appearance and presentation of digital images.
3. Character Styles – They manage the format of individual characters within a text string, and are usually imbedded within a paragraph style as a "sub-style".

The styles are an essential component of document automation. The automation approach we are using is dependent on processing and placing imported XML content streams. It would be hard to accept this approach as automation, if once content was loaded into the generated pages, one had to go back and format all of it manually.

However, this isn't the case. The individual XML tags in the content stream are mapped to individual styles. The styles are then applied as the XML content stream is imported and pages are generated. To facilitate mapping, avoid mistakes, and make life easier, the common recommendation is to give each style the same name as the XML tag with which it will be paired.

For document designs that will be reused and documents of similar design, all the ingredients exist for the creation of an automated document template: an XML content stream specification for variable content, a design blueprint, page masters with static content, and formatting styles matched with XML content tags. The only remaining step prior to contemplating a template is to test all components by actually generating the document print stream – our next topic.

Print Stream

PDF Generation

During the design preparation process, we've been working with a truncated content stream. It's truncated because, except for the data root and first record, all the remaining row elements were deleted from the imported XML content stream (Listing 10-1).

In practice, it is especially desirable for the truncated XML file to contain several rows (records). You can't test automation with one record; three or four are probably sufficient in this "learning" environment. Production environments may require more.

When all the design preparations are completed, the XML content stream is imported a second time into the data root XML element and loaded into the first document. White space deletion may be required using InDesign's StoryEditor.

When the first page is satisfactory, the content overflow can be loaded into the pointer and then dropped onto the next page with a mouse click while holding down the shift key on the keyboard. This second page and all remaining pages are automatically generated from the imported XML content stream. If you are thoroughly confused at this point, you may want to look at Maivald & Palmer (2008) or another source for more InDesign operational details.

```
<dataroot>
<row>
<name>Anthony</name>
<image href="file://Berries.jpg"/>
<location>North Side</location>
<offer>5% Discount</offer>
<product>Paving Stone</product>
<title>Mr.</title>
<first>Anthony</first>
<last>Able</last>
<street>27 Able Street</street>
<city>North Side</city>
<state>NY</state>
<zip>14533</zip>
<code>14533-27200803</code>
</row>
</dataroot>
```

Listing 10-1. A Truncated XML Content Stream

PDF Preparation

The generated pages make up the print stream. How the stream looks will depend on whether single pages or facing pages were selected at design start. Figures 10-2 and 10-3 (below) demonstrate the look of the two print streams.

The one characteristic the different print streams are liable to share is the existence of superfluous blank pages at the end of the content stream. These and other blank pages can be deleted. The temptation is to delete them as they sit in Adobe InDesign. Don't succumb! While this may work in some cases, it may also get pages out of sequence (especially facing or double sided pages), and that can be a real headache.

Make life easy on yourself! Generate an appropriate PDF file (i.e. a "high quality" or "press ready" PDF depending on the printer or target devise). Load the PDF into Adobe Acrobat Professional and select "View>Navigator Panel>Pages" from the menu bar. Delete blank pages using the panel's "Options" menu or right click the selected page to delete using a floating menu.

Figure 10-2 (next page) presents two variations of a non-facing page print stream. Both contain variable content. The first (far left) is a string of single-sided documents. The second is a string of double-sided (duplex) documents. The salient feature is a single static side that will print with each variable side. As a last bit of advise, start at the end of the print stream and work backwards. Delete trailing blank pages first, then delete any leading blank pages.

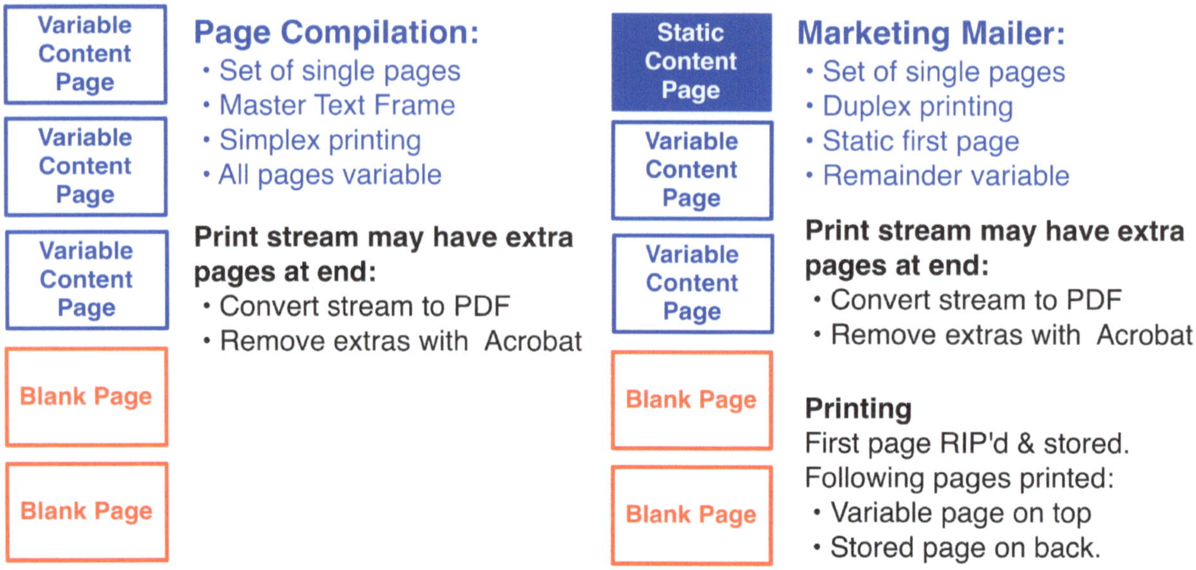

Figure 10-2. Print Stream from Single Page Spreads

The print streams illustrated by Figure 10-3 (below) were both designed with facing pages. Each document in "Marketing Mailer A" has a variable side and a static side. Each document in "Marketing Mailer B" has two variable sides. Beside the trailing blank pages, Mailer "A" has an out-of-sequence static page. First delete the blank trailers. Then click and drag that first static page next to the last variable page, making a properly completed pair.

Mailer "B" is easy to adjust – delete the trailing blank pages, and the the leading blank page. The bottom line is this: whatever the structure of the print stream, delete trailing blanks first, then leading blanks, and finally make any other necessary adjustments to clean up the PDF file.

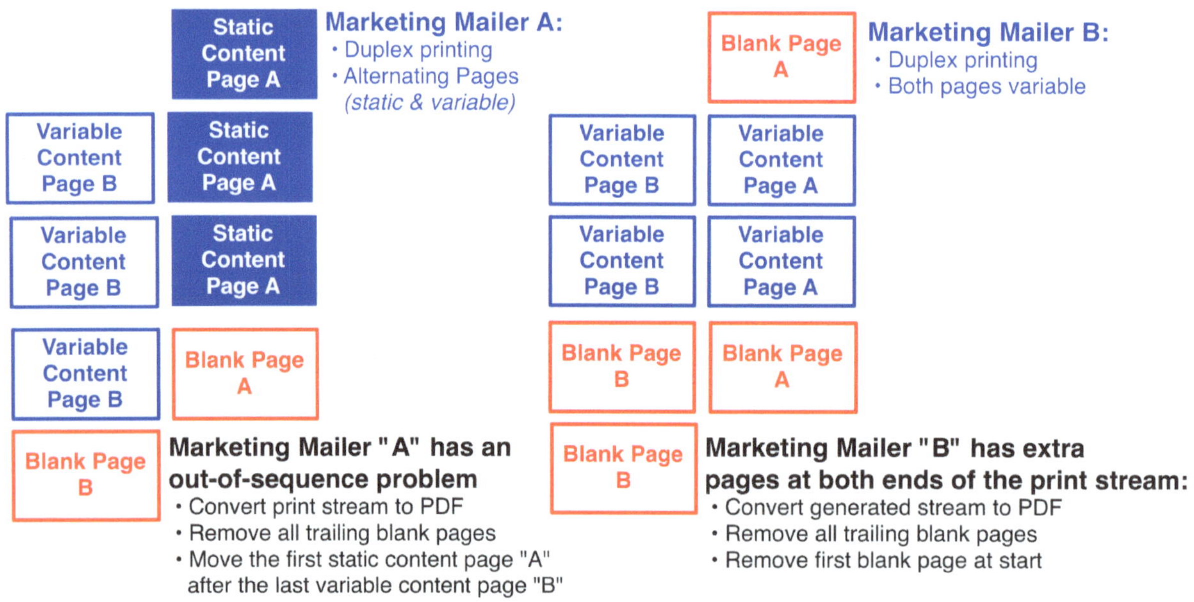

Figure 10-3. Print Stream from Facing Page Spreads

Recommended Reading

The "Classroom in a Book" series is excellent. An earlier edition, although for Adobe InDesign CS2, provided more than adequate reference for the purposes of this book. Newer editions are available for later versions of "Adobe Creative Suite". *Maivald & Palmer gets the reader into the detail with a series of very understandable explanations and very doable exercises. Powell is a good follow-on, because once a design process has been automated, it makes sense to make it a template for further use.*

Adobe. (2005). Adobe InDesign CS2 classroom in a book. Berkeley, CA: Peachpit Press.

Maivald, J. & Palmer, C. (2008). *A designer's guide to Adobe InDesign and XML.* Berkeley, CA: Adobe Press, Peachpit.

Powell, G. (2008). *Instant InDesign: designing templates for fast and efficient page layout.* Berkeley, CA: Adobe Press.

FOUR: Workflow

So far, we've been progressing in sequential order: first graphics and imagery, then data, and finally text and XML. That's fine for examining our series of technologies. However, how is work actually accomplished? In other words, what is the work flow? What is the work platform or infrastructure, and what is a basic toolkit? Given workflow, platform, and tools, how do we manage and maintain quality. Those are all questions that this unit will address.

11. An Iterative VCP Workflow

This first chapter talks about what is commonly viewed as the traditional approach to developing documents and the problems it presents when dealing with variable content. We then focus on an iterative alternative workflow, one that reflects how work is actually accomplished given the flexibility of modern development tools.

a. Rationale
b. Good Intentions
c. Depiction
d. Definition
e. Implications

12. Nothing Flashy – A Basic VCP Tool Kit

Moving right along, we look at a basic development environment and and an accompanying development toolkit. The chapter describes a limited collection of tools already alluded to in prior chapters to perform specific tasks or solve particular problemss.

a. Rationale
b. Development Platform
c. Development Environment
d. Self-study Suggestions

13. Intent, Performance, & Quality

Given all the flexibility, reiteration, interdependencies, and potential recursion, the need for planning, management, and quality assurance are significant factors that affect the successful execution and completion of variable document projects. That is what Chapter 13 is all about.

a. Practical Intentions
b. Setting Objectives
c. Identifying Requirements
d. Planning For Performance
e. Pursuing Quality
f. Supplementary Reading

11. An Iterative VCP Workflow

Discussion Points:

- ✓ Rationale
- ✓ Good Intentions
- ✓ Depiction
- ✓ Definition
- ✓ Implications

Rationale

Printing and publishing? It started with stone, then clay, followed by papyrus, vellum, parchment, and paper. Ink on paper started with pen and hand, and advanced to various kinds of printing presses set with carved wood letters, dies from carved molds, back to stone (lithography), hot metal, film, plates, and other variations on technology. They all had one significant characteristic.

Julius Caesar unknowingly identified that characteristic in 49 BCE with his utterance as his legions crossed the Rubicon river to seize the reigns of government in Rome, instigate a civil war, and launch an empire: "*Alea iacta est!*" – the die is cast, or figuratively, there is no turning back. For the early printer and the modern communications specialist (whatever the media}, the expression is both a figurative and literal play on words. Once hot metal is cast, film developed, plates etched, substrates inked, or communications expressed, it can be very costly (if even possible) to turn back. Consider the miss-spoken email!

For communications (especially printing), turning back can mean lost materials, lost labor, lost time, lost profits, lost opportunity, and possibly lost customers. So the process of going from blank slate to completed result engenders caution at each step. Statistically, the longer a project progresses from the starting point, the more costly it becomes to identify and correct an error, however inconsequential it might be. Most books on quality assurance will substantiate this.

To avoid the possibility of error, standardized processes (development methodologies in software and workflows in printing and graphic arts) embraced considerable caution (almost paranoia in some cases). Such approaches were often termed "waterfalls". "Process" became more rigorous, while demand for results accelerated. Waterfalls became bottlenecks as process took precedence over production, and work flows typically shrank to trickles.

Enter digital technology! Visual, interactive, point-and-click, digital capabilities spread across software development, graphic design, and printing. Document developers don't have to "cast a die" at every step. Print development is fluid, malleable, flexible, and (of course) variable. Creative and content-handling software provide more interaction, testing, and experimentation in designing and constructing graphic end products. However, traditional workflows particularly in printing, have remained relatively unchanged. Conclusion: we need to redefine workflows to fully exploit the capabilities of digital tools and techniques to meet increasing end-product requirements and possibilities.

Good Intentions

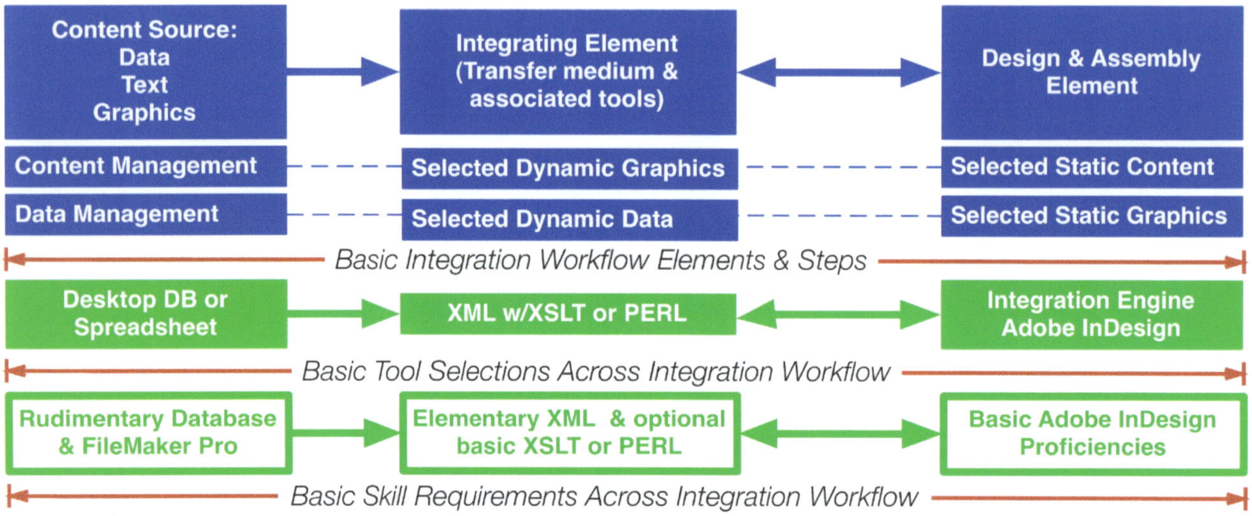

Figure 11-1. Basic VCP Integration Model

An obstacle to redefinition is confusion in our understanding of workflow. Take our chapter discussions up to now. They have been sequential: acquire content, clean it, organize it, prepare it, and integrate it. Take the basic integration model shown in Figure 11-1. The model suggests sequential processes, the tools change according to sequence, and applied skills also follow this same sequence.

What is frequently overlooked is the malleability of digital media. It's like potter's clay. Sure there are steps depending on the techniques employed. Sure there are different tools for different creative stages, and of course the tools and stages require different skills. But, the end product evolves on the potter's wheel – shaped, corrected, evaluated, and reshaped. The result isn't set until the first firing.

Trust & Hope

There are strong intentions by print and graphic practitioners to create quality results faster and at less expense. How does one do that? Well, a considerable amount of trust has been placed in accepted methods and work flows, bolstered by craft-based tradition, and the need to constrain expenses.

A considerable amount of trust has also been placed in technology in the hope that deficiencies or omissions in workflow, training, or know-how can be papered over with technological patches or "enhancements". Rather than redefining processes, identifying basic tool combinations, and promoting comprehensive training, the typical approach often relies on implementing new technological gimmickry within long-existing work patterns.

The belief is that existing practices and current skills will accommodate an increasingly sophisticated variety of digitally based end products simply by providing user enhancements, where "user friendly", really means "user-immune". Take the proverbial "black box", stuff it full, press the green button, and magically, exactly what's expected pops out the other side. Unfortunately, not necessarily so!

The strategy for moving from analog to digital print production becomes one of trying to fit digital technology into whatever methods and human resource capabilities are already available with as little inconvenience, expense, or upset as possible. Imagination and wishful thinking say that optimizing investments in increasingly capable digital tools simply means learning to click a few more buttons with the computer's mouse. That's essentially the hope!

Reality

The reality is something else! The user isn't dumb, just untrained. The training users receive is not basic or comprehensive enough to facilitate process conceptualization that subsequently translates into the ability to derive potentially new business solutions based on transferring learned knowledge to new problems (my opinion and experiences).

Teaching someone to bounce back and forth between panels or windows while pushing buttons in proper sequences does not provide the necessary working knowledge to progress from rote activity to creative solution design. Training examples centered on mail merge (I've experienced this) will most likely produce variations on the mail merge theme – over and over again! (If you don't believe me, just check your mail box.)

Change conjures up fear – images of uncertainty, discomfort, and possible failure. However, different materials and tools require different skills and techniques. For instance, modeling with clay is different from carving with wood. Non-pliant material is hard and resistant, requiring caution and care – the sequential waterfall workflow. Pliant material is forgiving, easily changed, and provides the opportunity to freely test and experiment – an iterative, interactive, visually-based workflow.

Depiction

For an effective workflow, we need to understand basic processes and how they fit together. We need to go beyond the simple three-step diagram depicted in Figure 11-1 and look at specifics. Figure 11-2 provides a detailed picture of the processes and relationships encompassed in the integration model.

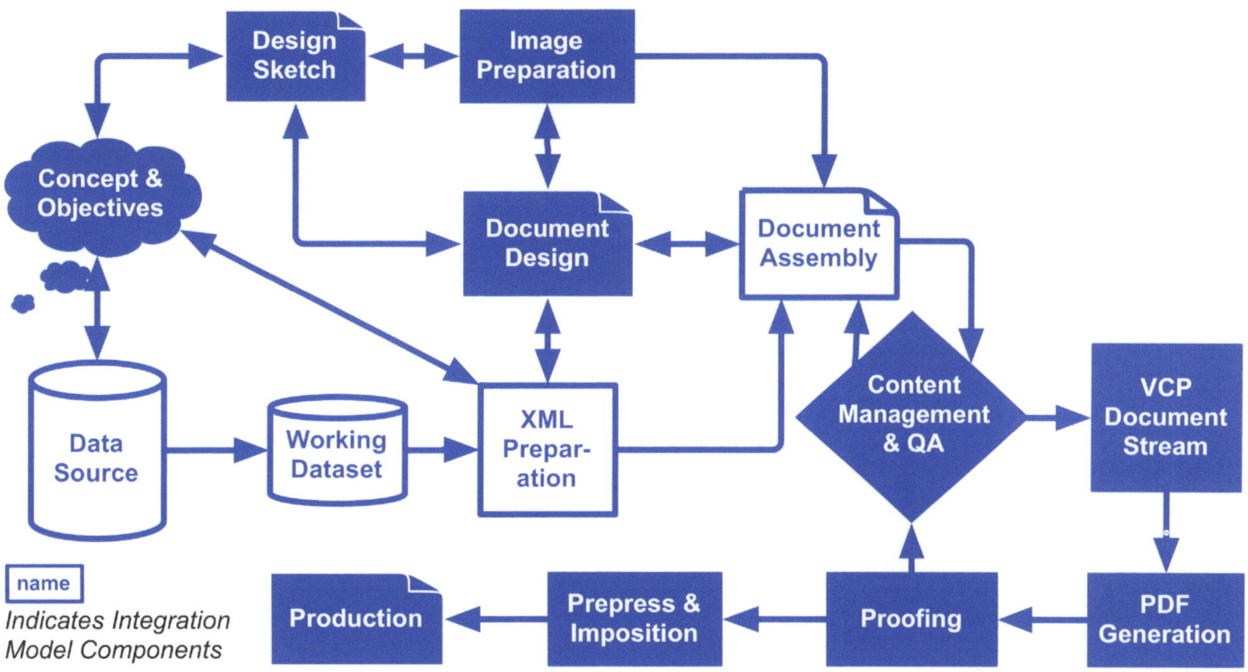

Figure 11-2. XML VCP System Workflow

This is an intricate workflow; in fact, almost too entangled for use as a standard process definition. It does, however, provide all the basic information needed to develop a digital workflow. Let's analyze the diagram without getting bogged down in arrows going every which direction:

Diagram Analysis

1. The different elements:
    a. Cloud – conceptualization and planning
    b. Drums – data stores
    c. Diamond – decision point
    d. Tabbed rectangle – document-related elements
    e. Squares or rectangles – processes
2. Element relationships:
    a. Lines with a single arrow at one end – sequential flows of information, content, or processes from one step to the next
    b. Lines with an arrow at each end – mutual dependencies, signifying recursive relationships between two or more components, where the results of activity or change carry forward and backward ramifications
3. The model starts with a concept and objectives, and ends with the production of intended documents
4. The handoff into the production environment essentially takes place with proofing and prepress steps
5. The model is a series of interconnected cycles, some running clockwise, some counterclockwise, and a couple of cycles capable of running either direction
6. In summary, the model conveys some significant implications for a workflow:
    a. There is continual forward movement from conception to production
    b. Progress is iterative and may affect multiple components of the model
    c. Multiple components can come into play simultaneously and finish within a short duration of each other
    d. Work in progress at whatever step must be accessible from any other step
    e. The predominant risk in the model is the potential for performing unnecessary iterations

Definition

Briefly, we have taken a conceptual digital development model (Figure 11-1) and restated it as a physical system constructed of intertwined processes (Figure 11-2). With that information, let's define an operational workflow for digital document and communications development. First we need to set expectations. What qualities does our digital workflow require?

Objectives

1. Generalization and, therefore, transferability:
    a. It should be appropriate for VCP and non-VCP document development
    b. It should be usable in differing system environments
    c. Consequently, it should be applicable for a variety of development tool sets
2. Repeatability – The workflow repeated on the same project will produce the same result, baring any changes to data, text, images, or design; used for another project, the workflow will also produce satisfactory results.

a. Therefore, project outcomes are predictable, and the activities themselves will also be predictable

b. Therefore, individual activities are identifiable and consequently measurable

c. So, the measurements themselves (referring to individual project activities) are recordable, trackable, and traceable, meaning that individual activities and sets of activities (procedures or processes) can be statistically analyzed and improved upon where necessary

Capabilities

Given these integral qualities, the workflow has use beyond serving only as a guideline for digital document development. What additional uses can the workflow provide?

- The basis for planning and estimating the development of document products, a framework for a work breakdown structure (WBS) at the activity level, and assistance in determining task interdependencies
- A foundation for document project tracking and analysis to support project budgeting and forecasting
- The basis for defining document development standards and procedures, and the application of quality assurance practices during actual development activities
- The fundamental starting point for document development training – especially in regard to VCP
- A sales tool that customer representatives and developers can be easily explained to customers or clients

Development described in the system workflow (Figure 11-2) consists of various process and specialized sub-processes:

- Objectives and project inception
- Project requirements definition (including both business and production requirements)
- Content management project setup
- Initial document designs
- Image preparation (incorporates a specialized sub-workflow) starting with project digital asset identification
- Data acquisition and data preparation (incorporates a specialized sub-workflow)
- XML preparation (incorporates a specialized sub-workflow)
- Design integration preparation
- Document assembly – physically integrating data, text, and digital images (packaged as an XML content stream) into the prepared document layouts and then generating the resulting print streams
- Quality assurance – problem identification, resolution, and production release regulation
- Production (printing and finishing)

Implications

As itemized, there are a considerable number of activities involved in digital document development. So how do all these processes fit together, and how is the result going to be any less complex than the system workflow? Figure 11-3 (next page) illustrates a digital document development workflow particularly, but not exclusively, suited to variable content projects. XML-based techniques can easily support development of both variable and static content document development. This workflow could just as easily apply to the digital development activities for computer-to-plate (CtP) processing where documents will be printed on an offset rather than digital press.

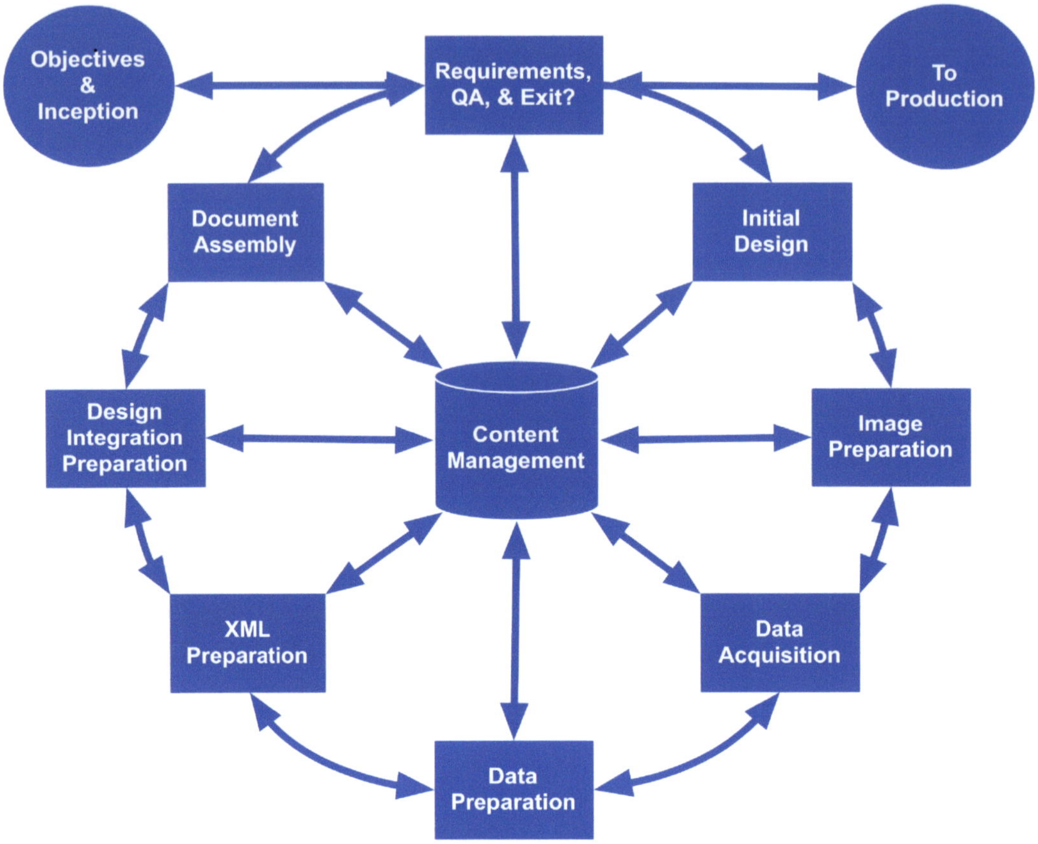

Figure 11-3. XML VCP Iterative Workflow

Figure 11-3 portrays how digital development is actually accomplished. Yes, there is a sequential aspect to the work, but to say that, "first do step A, then step B, followed by step C", is not realistic. We may start with step A, but in progressing to step C from B, we may find that we need to return to A momentarily. Or we may go through a complete development cycle, and realize on the last step that we need to do another iteration to correct or add a feature.

Some portions of work will be quite systematic. Work will effectively progress around the workflow cycle, but it won't always be as methodical as we might imagine. In fact, it may not be very pretty. Figure 11-4 (next Page) demonstrates various potential activities across development stages (dashed lines) together with commonly expected activities between successive work steps (solid lines). Why does work actually play out like this?

First, fundamental development still has its complications. Keeping track of all components that will comprise a document while making sure they all fit together in chosen locations with the desired appearance requires considerable attention to detail. Processes related to variability and automated document generation make more demands.

Second, given the complexity level, perfection is hard to achieve the first time out of the gate. People make mistakes, overlook small details, or miss problems that are "hiding in plain sight". Consequently, there will be rework. A repair in one place may require adjustments in another – especially for co-dependent work elements.

And third, the design and development tools used in creating digital documents – both variable and static – provide extensive flexibility that helps facilitate recovery from errors, oversights, and exception conditions. As a result, the development workflow must also accommodate this same flexibility. Because of the recursive nature of some development steps, some specialized work may appear rather random and haphazard.

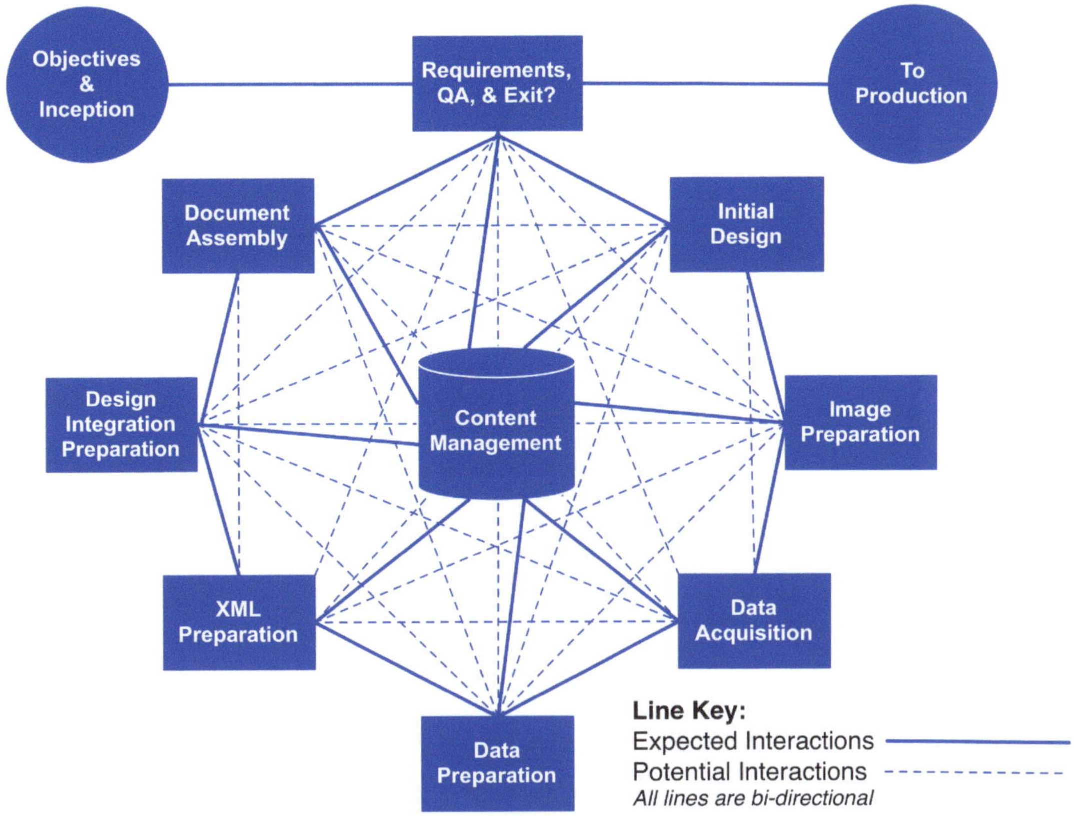

Figure 11-4. Workflow Interactivity

Figure 11-4 depicts how the different work components interact during the development effort. Considering this and our discussion around Figure 11-3, we can use this workflow to make a number of assumptions about managing digital assets, staffing, and the potential affects on the structure and inter-workings of organizations doing this kind of work:

1. Content management is the heart of digital document development:
   - Single document projects require management and version control to preclude duplicated efforts and mix-ups with source material, work in progress, and finished goods.
   - Multiple document projects increase digital asset management needs for text, data, and images geometrically as the number of documents under development increases.
2. Project staffing:
   - *Single document projects generally require only a single person possessing design, data, and XML manipulation skills. Otherwise, two people may be required – one for design and one for data and XML handling*
   - *Multiple document projects may require more developers depending on the number and complexity of the documents. Key roles will be a digital asset librarian, a lead designer, and potentially a specialist for data and XML.*
3. Organizational ramifications:
   - *The combination of digital workflow and digital tools mitigates any need for departmental boundaries that may have been established in defining creative, pre-media, prepress, and production activities.*
   - *In considering the entire workflow (start to finish), a very close working relationship needs to exist between the development effort, sales (the customer interface), and digital press operations (physical product generation)*

- *The digital workflow requires customer availability (at minimum) during the project, may necessitate involvement of customer resources during initial project activity (requirements specification, data and digital image acquisition, and feasibility determination), and warrant customer participation during requirements compliance, proofing, and final product acceptance*

Keeping the workflow discussion in mind, in the next chapter we will examine a basic VCP development tool kit and the limited set of technical skills required to utilize it. (*The reader should probably have a pretty good idea by now of what will be discussed.*) From there, we will look at some primary factors in successfully completing a digital document project – setting objectives, specifying requirements, planning and managing the project, and assuring quality results.

12. Nothing Flashy – A Basic VCP Tool Kit

Discussion Points:

- ✓ Rationale
- ✓ Development Platform
- ✓ Development Environment
- ✓ Self-study Suggestions

Rationale

Up until now, we have primarily been talking about processes, techniques, and a small core of basic tools that provide all the capabilities needed to produce VCP documents. Our overriding objective has been to stay on the fundamentals – keeping it simple – while doing more with less. That is still true.

Now we need to expand our discussion into a further area, pulling everything together on a functional platform environment, followed by some thoughts about putting it all to work. We'll still keep it simple. We'll also avoid getting too explicit, because discussion details vary on a case-by-case basis – usually around scale, quantity, technology, and cost. This may seem vague right now, but you'll see my point as we proceed through the chapter.

Development Platform

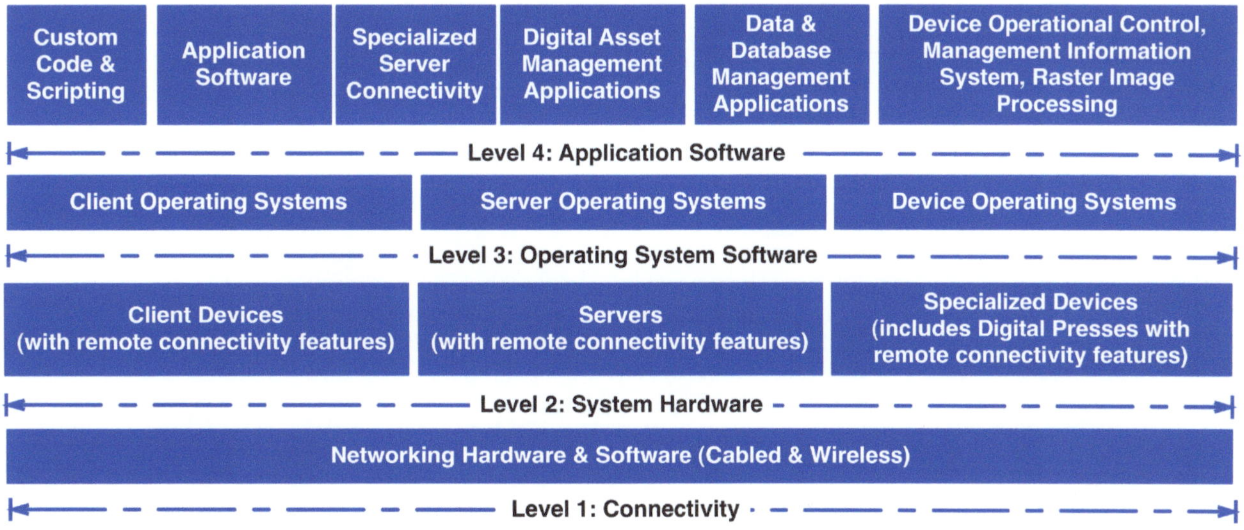

Figure 12-1. Basic System Architecture

Starting with the host platform (or platforms), the development system may range from a single high-powered laptop to a multi-server multi-client system. That single laptop may be a PC or Mac. The multi-tiered system could conceivably be a mixture. This discussion is based on research using Apple's OS X 10.5 and Microsoft's Windows XP Professional. Whatever the brand or version, at a conceptual level, system architecture is relatively uniform, and although rapidly transforming with new capabilities and approaches, it remains a relatively stable framework. Figure 12-1 (previous page) illustrates what we're talking about.

System architecture is like a four-layer cake. However, it's not all chocolate. Each layer has different ingredients, consistency, and flavor. In other words, each level has different combinations of hardware and software, different functionality, and different intended purposes. Let's start at the foundation:

Connectivity (Level 1)

The initial temptation says, "let's get the computers up and running and then worry about connecting them together." If you've ever had a house built or watched the construction of a housing tract, you know that some of the very first activities are land preparation and installation of basic sewer, water, and electrical services – possibly even phone and cable services. Connectivity! The reason?

- It establishes service connection standards and requirements
- It establishes the locations where all the units will be placed
- It simplifies installation of the primary backbone for service (distribution and collection)
- It simplifies establishing connections between the service backbone and individual units as they come online

Doesn't that sound the same as establishing a computer network. Conceptually, it is! How we intend to develop document products, and how we are going to integrate that development effort with other organization activities, internal or external to the workflow or enterprise, depends on what capabilities are needed to effectively connect at the right times, places, band width, and durations.

System Hardware (Level 2)

Computer hardware refers to "all the usual suspects": memory, processors, disk storage, displays, graphic processors, and more. The list can be very extensive or quite truncated, depending on how we're going to develop documents. Development could range from a single person using a fully-featured laptop computer to multiple development teams accessing very large client-server systems. Whatever the case, sage advice says to get the newest, biggest, and fastest devices affordable. The application of this adage is contingent on a host of factors. When all is said and done, the keys are:

- Speed (processor count and rating, memory speed, and disk speed)
- Capacity (memory and storage)
- Graphics (color, speed, and visual acuity)
- Sound (input, sampling, wave manipulation, output clarity, and video coordination)
- Expandability and flexibility (the ability to add, change, or enhance subcomponents)
- Portability (may be more a presumed need than a real one)

For a laptop, since expandability is limited, it means "maxing it out" at time of purchase. Doing this is not inexpensive, but it will considerably extend the usable life of the computer, mitigating some initial expenses. Non-portable computer configurations are very flexible and expandable. They are upgradable at the component level as business needs, product or service offerings, and customer requirements evolve. One direct benefit may be easier balancing of operating expenditures, technology purchases, and revenue potential.

Operating System Software (Level 3)

The operating system manages the computer's resources and provides the basic user interface. It also participates in managing communications. Basic operating system functionality is usually overlaid with increasingly capable graphical user interfaces (GUIs). This insulates the user from typing text commands through a terminal screen via the keyboard. Rather, users have an event-driven interface that provides "point and click", "drag and drop" functionality.

Computers running different operating systems can co-exist within the same networking system and cooperate effectively. The issue is not so much "mixing things up" (for instance, running two or three different operating systems) and therefore not being able get work done. Work can be accomplished just fine! The issue with running multiple operating systems can actually, and quite literally, be "mixing things up" in regard to system maintenance, problem resolution, cross training, operational coverage, and simply remembering particular operating system specifics and shortcuts.

Repeating a consistent thread in our discussions, make it easy on yourself and simplify. Standardize on a single operating system; it's much easier to service and maintain. That doesn't mean the entire enterprise has to run the same system. It does imply, that life may be easier at departmental and workgroup levels with standardization. The exception is when a particular job requires application software not compatible with the predominant operating system.

Application Software (Level 4)

A similar approach can be applied to application software – keep it simple. That doesn't mean using anemic applications. It means assembling a robust toolkit that minimizes overlaps in functionality. Such a toolkit not only simplifies training, but also constrains unneeded expense. Toolkit details are presented next.

Development Environment

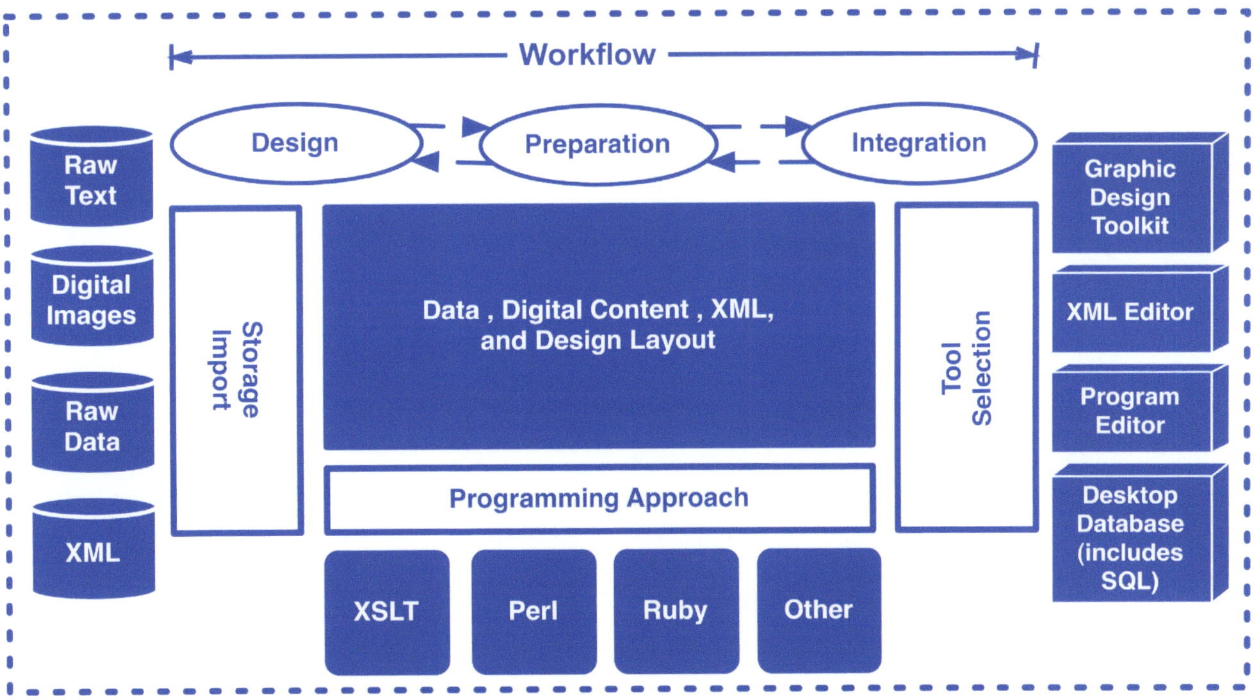

Figure 12-2. Basic VCP Development Environment

Figure 12-2 (prior page) presents an abstraction of a digital workshop for document development. This development environment is hosted by the the system platform (Figure 12-1). This is where processes, tools, techniques, and content are brought together to forge document solutions. Let's examine it, starting at the top and moving clockwise around the illustration.

Development Interplay

The three ovals across the top (design, preparation, and integration) are an amalgamated extrapolation from the iterative VCP workflow discussed earlier. They represent visible efforts:

- "Design" incorporates initial document construction and fine adjustments to the layout, which are usually discovered (or uncovered) during content preparation and integration activities.
- "Preparation" is the combination of activities that prepare the different types of content (graphics, data, XML, and text) for integration with the document layout.
- "Integration" activities encompass taking care of all the final details for successfully bringing together document design and content elements by executing the integration program logic, which may require three steps:
    - An abbreviated initial setup run,
    - An abbreviated correction and test run,
    - And a final full-length production run to conclude the integration effort.

These three categories of activities usually run concurrently and complete simultaneously. They are closely interrelated and dependent upon each other, hence the dotted bidirectional connections between them, and their joint application to the core of the development workshop.

Tool Selection

One goal with regard to development tools is optimizing the variety of document solutions that can be produced while minimizing the number of tools needed to accomplish their production. The other goal is the optimization of efficiency and effectiveness throughout the development process.

Mathematicians and operations researchers in the crowd might view this as a linear programming problem. However, "trial and error" shows that a potentially large variety of documents can be effectively and efficiently produced with four tool types shown along the right side of the illustration:

- A graphic design suite such as Adobe's "Creative Suite"
- An XML editor such as SyncRO Soft's "<oXygen/> XML Editor"
- A programmer's editor, which provides advantages over a simple text editor (although the simple text editor is workable)
- A desktop database application such as FileMaker Pro or Microsoft's "Access". (A spreadsheet program like Microsoft Excel, is a handy supplement, but does not take the place of the desktop DBMS.)

The reduced tool set offers some attractive advantages, especially for small businesses, new entrants into digital variable solution delivery, and those who want to specialize in low volume high customization solutions.

- Smaller hardware and software acquisition costs during startup and initial operations
- Lower ongoing operating and maintenance costs
- Narrower training and skill maintenance needs for new and existing VCP developers

What remains debatable is whether the smaller skill requirements simplify finding skilled staff, or reduce ongoing payroll expenditures. "Less" does not necessarily tally up to "more" in every case. The smaller number of tools does not necessarily imply the need for less expertise. It may only facilitate gains in expertise at a quicker pace, having less subject matter to study. Otherwise, it probably implies the need for greater expertise in the few areas involved. Investing in more skilled practitioners may likely produce higher returns on investment, increased sales, and a better profits. So there are trade-offs to consider.

Programming Approach

Moving to the bottom of the illustration, the development environment needs a facility to work with text-based files, whether raw text, data files, or XML. If textual content is XML, then either an XSLT approach (via the XML editor) or a procedural approach with a text-oriented programming language such as Perl or Ruby. Perl and Ruby are also appropriate for raw text and data files.

Why select Perl or Ruby? Most VCP processing will involve ad hoc data file or XML manipulation, as opposed to creating fixed logic applications, GUIs, or code libraries. Therefore, compiled programming languages like C, C++, Objective-C, Java, and C# are liable to introduce complexities, such as recurring recompiles every time the developer makes a program change or correction. VisualBasic and Javascript raise their own obstacles.

Because VCP jobs are apt to involve different sets of data that require different processing, each job requires a high degree of flexibility. Programs written in an interpreted language like Perl and Ruby are converted into machine-executable instructions as the interpreter reads the written program. This means that every time the program is run, it is interpreted (essentially compiled at runtime) without human involvement. Because interpreted programs can be modified and "run on the fly" they are very flexible and capable of rapid turnaround.

This makes them ideal for VCP jobs. Additionally, both Perl and Ruby are extremely well suited for text processing (the primary need when manipulating data or raw text for VCP). Any program logic that can be reused with modification can be retained as programming templates or sets of functions that can be included in new program or copied and customized prior to running.

Both Perl and Ruby have other advantages. Training is relatively straightforward. A strong open source developer community supports both languages. Documentation and tutorials are plentiful. And both languages are cross-platform capable, meaning for example, the same program can run on Mac OS X, PC Windows, Linux, or Unix. And both are cross-media capable, meaning they can also be used on the Web.

Storage Import

Along the left-hand side of Figure 12-2 are likely sources of unprepared content that will be used in various VCP projects. Raw text, unprocessed digital images, unrefined data files, and unreviewed or unfiltered XML are the usual types of encountered content. These assets initially enter the work environment individually or in collections of raw material for staging to predetermined projects for further processing.

Action Central

The diagram's core represents the four primary activities that occur in the work environment:

1. Initial analysis and staging of content prior to performing more extensive processing steps,
2. Use of application software to assist in layout design and content preparation processes,
3. Development and use of ad hoc programs to automate processing of text, data files, and XML,
4. And integration setup and testing of prepared design components, document generation, PDF conversion, and print stream fine tuning.

Self-study Suggestions

This short list is for those who desire a more comprehensive exposure to the technologies we've been discussing. These books provide a very good way to acquire skills through self-study. They are all loaded with examples. Some provide exercises to gain hands-on experience with the topics being presented. In addition, there are considerable resources for each of these topics on the Web, including reference material, tutorials, and online technical support.

Desktop Database with Cross-platform Capability

Coffey, G. & Prosser, S. (2005). *FileMaker Pro: the missing manual.* Sebastopol, CA: O'Reilly Media, Inc.

Ruby Programming Language – object-oriented approach

Cooper, P. (2007). *Beginning ruby from novice to professional.* Berkeley, CA: Apress.

Fitzgerald, M. (2007). *Learning ruby.* Sebastopol, CA: O'Reilly Media, Inc.

Adobe InDesign

Maivald, J. & Palmer, C. (2008). *A designer's guide to Adobe InDesign and XML.* Berkeley, CA: Adobe Press, Peachpit.

Powell, G. (2007). *Instant InDesign: designing templates for fast and efficient page layout.* Berkeley, CA: Adobe Press.

Perl Programming Language – procedural approach

Quigley, E. (2008). *Perl by example, fourth edition.* Boston, MA: Pearson Education, Inc. and Prentice Hall.

Schwartz, R., Phoenix, T., & Foy, B. (2005). *Learning Perl, fourth edition.* Sebastopol, CA: O'Reilly Media, Inc.

XML

Wagner, R. & Mansfield, R. (2003). *XML all-in-one desk reference for dummies.* NY: Wiley Publishing, Inc.

13. Intent, Performance, & Quality

Discussion Points:

- ✓ Practical Intentions
- ✓ Setting Objectives
- ✓ Identifying Requirements
- ✓ Planning For Performance
- ✓ Pursuing Quality
- ✓ Supplementary Reading

Practical Intentions

The purpose of this chapter is to consider common factors that may influence how successfully a VCP project is planned, managed, and executed. This is not a "how to" chapter. It is a "pause and reflect" chapter. Many of the ideas discussed are commonly associated with project management. Others deal with issues related to quality assurance.

This chapter does not pretend to "instruct" about the best way to do project management or quality assurance. There are whole books written in considerable numbers on these two subjects. What fool would presume to cover both topics in a single chapter? Besides, there are many styles and approaches to either discipline, some favored on a personal level, others on an organizational level.

So, the ideas presented will usually be expressed as sets of questions in the form of checklists. They are relevant to most projects, but particularly to VCP and digital printing. However, these are not exhaustive lists of questions (although the reader may feel rather exhausted after reviewing a few of them). It's left to the reader to pick what's helpful and what's not, given their particular circumstances.

Setting Objectives

The first step in promoting project quality, operational efficiency, and optimal effectivity is to define objectives, quantify them for tracking purposes, then measure and evaluate results from the delivered products and services. Figure 13-1 (next page) summarizes objectives-setting. It illustrates key factors and their interrelationships.

Figure 13-1 is the starting point for three sets of small checklists intended to assist in initially defining project objectives, and in making reasonably certain that all bases are covered. Obviously, every circumstance can't be addressed, but if these first checklists stimulate further thinking, then they will have met their own purposes.

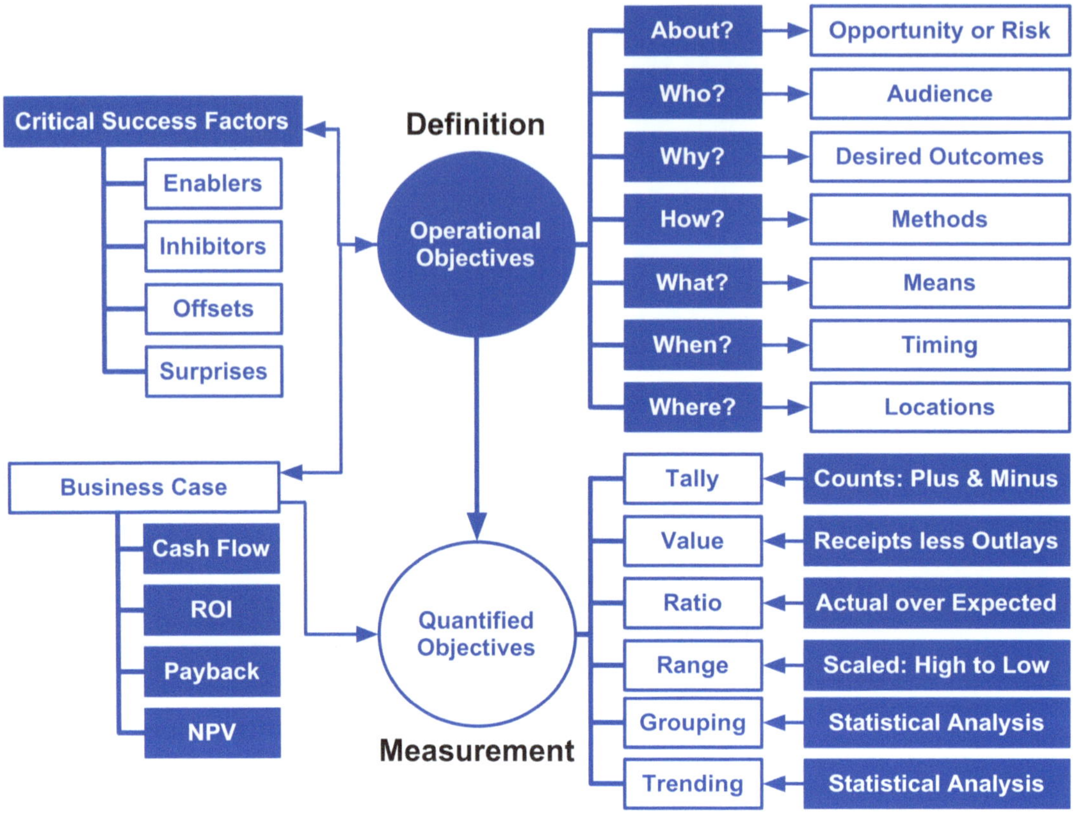

Figure 13-1. Setting Objectives

Checklist 13-1 provides additional depth to Figure 13-1. The purpose of Checklist 13-1 is to step back for an initial perspective on internal and external organizational factors that are central to defining a project. By the way, this approach would hold true for most kinds of projects.

Checklist 13-1. Outward & Inward Perspectives

- ❒ What is the focus of the project – who is the intended audience?
- ❒ Is the project aimed at businesses and individuals outside the organization?
- ❒ Will the audience be segmented, and if so, according to what criteria?
- ❒ Is the project aimed at accomplishing something within the organization (for example, some sort of employee communication)?
- ❒ Is the project aimed at realizing both internal and external goals, such as enticing customers with a sale while relieving overstocked inventory at particular locations?

End of Checklist

Checklist 13-2 (next page) switches from an organizational viewpoint (concept) to focus directly on factors affecting the conduct of the project (execution) and the anticipated results (consequences).

Checklist 13-2. Motive, Means, & Opportunity

- ❐ What is the need or opportunity, and how will the project address it?
- ❐ What specifically are the desired results?
- ❐ When the project is completed, what will be different and what will be the same?
- ❐ What resources will be needed for the project?
- ❐ What approach or methods will be used to prepare and execute the project?
- ❐ When will the project take place and in what sequence of events?
- ❐ Where will the project be executed and what locations will be involved?
- ❐ What are the risks, the potential benefits, and does one outweigh the other?
- ❐ Who or what are potential enablers and inhibitors and do they offset each other?
- ❐ Are there any potential surprises?
- ❐ Is there a business case, and if not, then why not?
- ❐ What factors and analysis methods are included in the business case?

End of Checklist

Now, as the project sits on the brink of being executed, Checklist 13-3 essentially asks how we will know when we have arrived at the project destination (completion), whether the result is what we intended (effectiveness), and how well we performed in making the trip (efficiency).

Checklist 13-3. Measuring Achievement

- ❐ What will signify that objectives were accomplished?
- ❐ How will individual results be measured?
- ❐ Will results include both quantitative and qualitative indicators?
- ❐ What results will signify the validity of the business case?
- ❐ What aggregates, limits, or proportions will be used to evaluate achievement levels?
- ❐ Will any anecdotal or testimonial accounts be included in an evaluation?

End of Checklist

Building Traceability

Once objectives are established, they provide the basis for identifying requirement details and ultimately establishing a project execution plan. The interrelationship of the three levels (objectives, requirements, and planning) creates a bidirectional audit trail, meaning that individual statements or details can be traced from point of origin to point of application, or from application back to origin. Figure 13-2 (next page) demonstrates how traceability promotes congruence and clarity.

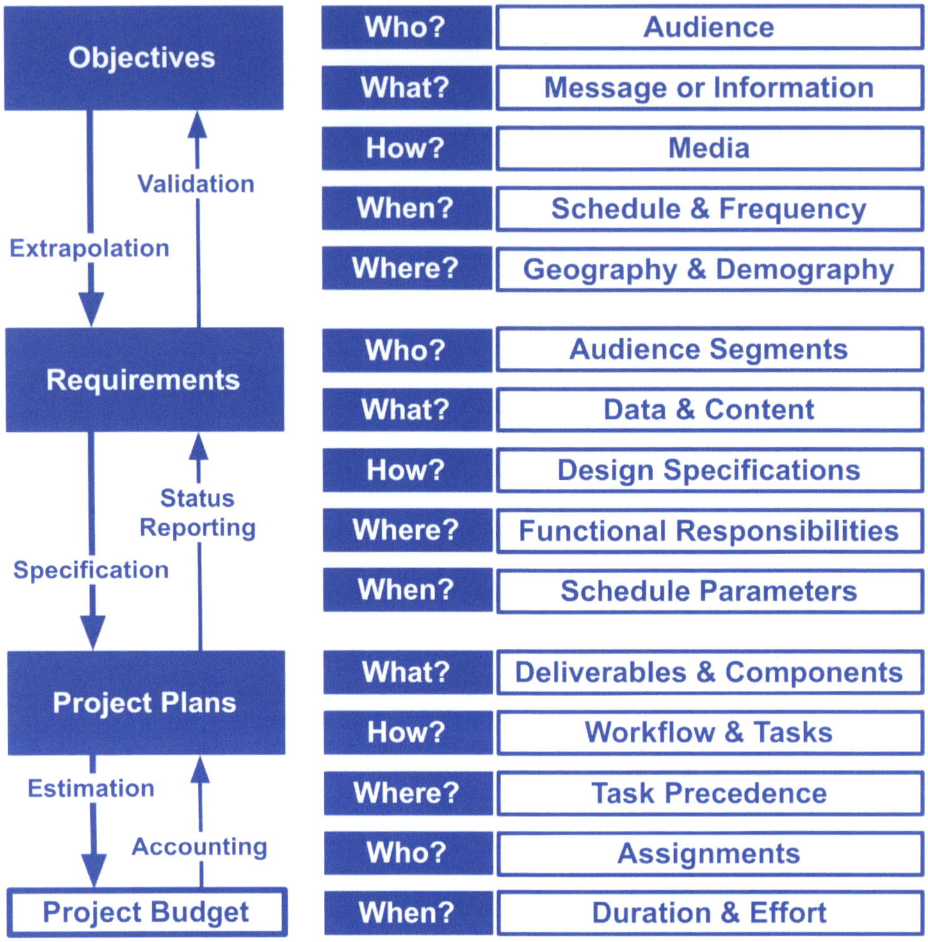

Figure 13-2. Traceability in Project Planning & Budgeting

Identifying Requirements

Let's move into requirements identification, continuing to look at Figure 13-2. The basic questions that helped define objectives remain largely the same for requirements, but elicit more detailed and specific responses. Requirements identification isn't a one-to-one process where a single objective statement is simply elaborated upon with more details.

Rather, it is both a decomposition and an extrapolation process where each objective is logically subdivided into multiple categories of results with increasing elaboration and detail. While the checklists below don't cover every possible circumstance, they do provide a flavor for the kinds of questions that may be posed during this process.

The easiest way to look at requirements is to break them (divide and conquer) into at least four main areas that ultimately combine to produce final products and attendant results: Design Requirements (Checklist 13-4), Digital Asset or imaging Requirements (Checklist 13-5), Data Requirements (Checklist 13-6), and Production Requirements (Checklist 13-7). (All four are listed on the following three pages.)

Checklist 13-4. Design Requirements

- ❒ If the document is a mail piece, does the design comprehend postal sizes, dimensional standards, and machinability constraints or restrictions?
- ❒ Does the design comprehend digital printing substrate requirements: weight, material, and size?
- ❒ Does the design comprehend digital printing toner coverage limitations?
- ❒ Does the design comprehend the aftereffects of finishing (particularly folding, creasing, and scoring) with regard to substrate grain direction and potential toner cracking or scratching?
- ❒ Does the design comprehend space allowances to accommodate cutting, drilling, and selected binding?
- ❒ Does color selection take into account potential digital printing tonal variation across large single-color spaces within a document or across multiple documents?
- ❒ Do multi-page designs for duplex printing accommodate pagination, adequate margins, and gutters?
- ❒ Do single sheet duplex document designs facilitate accurate alignment and substrate utilization?

End of Checklist

Checklist 13-5. Digital Asset Requirements

- ❒ What digital assets are needed for the project and how are the digital assets going to be used?
- ❒ Is there a preliminary digital asset inventory and what digital assets are currently available?
- ❒ How will digital assets be managed for the project?
- ❒ Will digital assets be acquired from an outside source, at what cost, on what medium, and when?
- ❒ What are the digital asset file types, their individual compression ratios, file sizes, and dimensions?
- ❒ What is the quality of individual digital assets?
- ❒ Do the digital assets have embedded color profiles, and if so, what are they?
- ❒ How will the source color profile be converted to the target color profile?
- ❒ How will the digital assets be used (as information or strictly decoratively)?
- ❒ Will all digital assets used in the project be included in the project archive?
- ❒ Will digital assets be retained after production for reuse, or can they be cleared from work-in-process?
- ❒ If project digital assets are retained in work-in-process for reuse, how soon will that work begin?
- ❒ What will be the disposition of project-specific digital asset originals housed in the content collection?

End of Checklist

Checklist 13-6. Data Requirements

- ❒ What data is needed to support the project?
- ❒ Has a preliminary data analysis been made?
- ❒ What data is currently available and in what file format?
- ❒ Will the data be furnished or acquired from an outside source, at what cost, on what medium, in what format, and when?
- ❒ Is data from single or multiple sources, will it need extraction, when, and how?
- ❒ Is the data organized in single or multiple data sets?
- ❒ What is the quality of the data?
- ❒ Has any data preparation or cleansing already been done, and if so, what?
- ❒ If edited and cleansed, will the corrected data be used to update source data?
- ❒ What file format will be used for working data and how will it be converted to the desired format?
- ❒ How is the data going to be used – for data mining, addressing, customization and personalization, some combination of these, all of them, or something else?
- ❒ How will data be managed during the project?
- ❒ What facility or application will provide intermediate data storage?
- ❒ Will all data used in the project be included in the project archive?
- ❒ Will data be retained after production for reuse in followup or new work, or can it be cleared from work-in-process?
- ❒ If project data is retained in work-in-process for followup work, how soon will the work commence?
- ❒ What will be the disposition of project data held in unworked (raw) data storage?
- ❒ What decisions have been made regarding document recipients, given available data?
- ❒ Can "house-holding" or some other approach be used for multiple recipients at the same location?

End of Checklist

Note that text is being treated here as data, since composition, editing, XML preparation (if applicable), storage, and access can be handled by technical processes akin to data handling.

Checklist 13-7. Production Requirements

- ❒ When is production required or anticipated, and has it been scheduled?
- ❒ What digital printing technology will be used – for instance electrostatic, inkjet, or something else?
- ❒ What is the selected substrate in terms of cut sheet or continuous feed, size, weight, brightness, finishing, and grain?
- ❒ Is the selected substrate on hand, easily available by order, or special order that may require additional shipping time and handling expense?
- ❒ Are special toner colors required in addition to standard 4-color printing requirements (CMYK – cyan, magenta, yellow, and black)?
- ❒ Is adequate toner available (including any special colors) or must any be ordered, requiring additional shipping time and handling expense?

- ❒ If a document is a mail piece, will the toner be clear-coated to help prevent scratching and other processing wear from postal equipment?
- ❒ If not mail pieces, is clear-coating desirable anyway?
- ❒ Will documents require additional processing such as finishing, binding, scoring, folding, trimming, drilling, collating, stuffing, or something else, and is the added technology or services available?
- ❒ If a document is to be bound, has a particular style of binding been selected, and are necessary supplies, technology, and expertise available?
- ❒ What is the plan for proofing (soft, hard, both), and what technology and equipment are needed?
- ❒ Has an impositioning scheme been worked out, and if so, what?

End of Checklist

Reflection – Are Objectives Feasible?

Once requirements have been classified and itemized, there should be enough understanding to determine if stated objectives are feasible, particularly with regard to available data and digital assets. ***This is not feasibility from the standpoint of projected costs and revenues derived from the business case.***

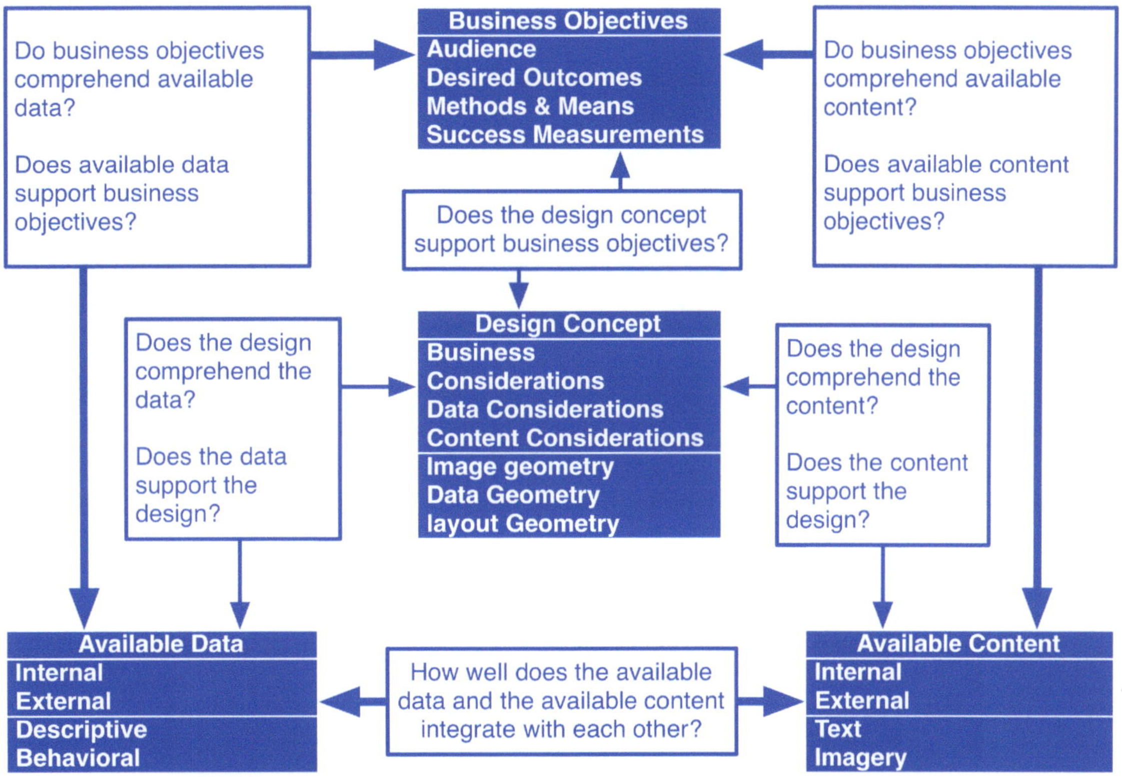

Figure 13-3. Reflection – Objectives, Digital Assets, & Data

Considering that the success of a VCP project is primarily driven by the quality of available data and the relationship between data and imagery, the implication is that the achievability of stated objectives has similar dependencies. Figure 13-3 depicts these different interrelationships and the questions they pose. The overarching question posed by Figure 13-3 is this: "***Do stated objectives comprehend available data and graphics, and do the available data and graphics support stated objectives?***" In other words, are the objectives, and therefore the project, feasible?

Planning for Performance

A project presents a problem. The natural response is to ask how we are going to solve it. Similarly, doing a project means activity, so the next question is, unsurprisingly, what are we going to do, followed shortly by who's going to do it. Obeying the project mantra, "on time on budget", great haste and energy is expended in doing the project. As time passes the proverbial eighty-twenty rule kicks in: eighty percent of estimated effort, cost, and elapsed time have been used to move the project twenty percent toward completion. That is not exemplary performance, but it may be typical.

Where do we start laying out the plans for a project? There are different approaches and preferences: concentrating on tasks, people, skills, processes, or end products. VCP projects are intended to produce results – end products or deliverables. Therefore, the best place to begin defining a VCP project plan is to start with deliverables (what the project is going to deliver) by defining a manufacturing bill of materials (BOM).

Everyone has probably come face to face with a BOM at some point: after purchasing a lawnmower or snow blower, or putting together a gas grill, a garden wagon, a child's bicycle, or a backyard swing set. You know what it is; that itemized hierarchical list of all the major components, subcomponents, screws, nuts, bolts, washers, chains, and wires that make up whatever it is that you're assembling or servicing.

Many just don't think of software and graphics projects like VCP in those terms – parts and pieces; but VCP is a component assembly process from start to finish. Now, we are not going to demonstrate planning a project step by step, but we are going to provide a checklist (as you might have suspected) of appropriate questions in an order suitable for producing a performance plan aimed at achieving desired results as well as facilitating process improvement.

Checklist 13-8. Planning Questions for Order of Execution

- ❒ What finished products (major deliverables) will the project produce?
- ❒ For each product, what are its subcomponents?
- ❒ For each product, what tasks are necessary to create and integrate subcomponents into a finished deliverable or end-product?
- ❒ What is the precedence (the order in which tasks will be done) for each task?
  - ✓ What tasks can start first (they are not dependent on any other task finishing)?
  - ✓ What tasks are start-to-start relationships (they can start at the same time and work in parallel)?
  - ✓ What tasks have a start-to-finish dependency (one must finish before the other can start)?
- ❒ What skills and tools are needed for each task?
- ❒ Who is available with the appropriate skills to do the task?
- ❒ Considering staff availability, do any dependencies require adjustment to make accommodation?
- ❒ How much effort (work) is needed to complete each task?
- ❒ For each task, how much elapsed time (duration) is needed to accommodate the effort?
- ❒ Given effort required and rate for the assigned staff member(s), what is the cost of each task?

End of Checklist

While this set of questions may seem simplistic, we are not building an aircraft carrier, or going to Mars. The result of this question/answer exercise should help produce a project plan that looks something like the model shown in Figure 13-4 (next page).

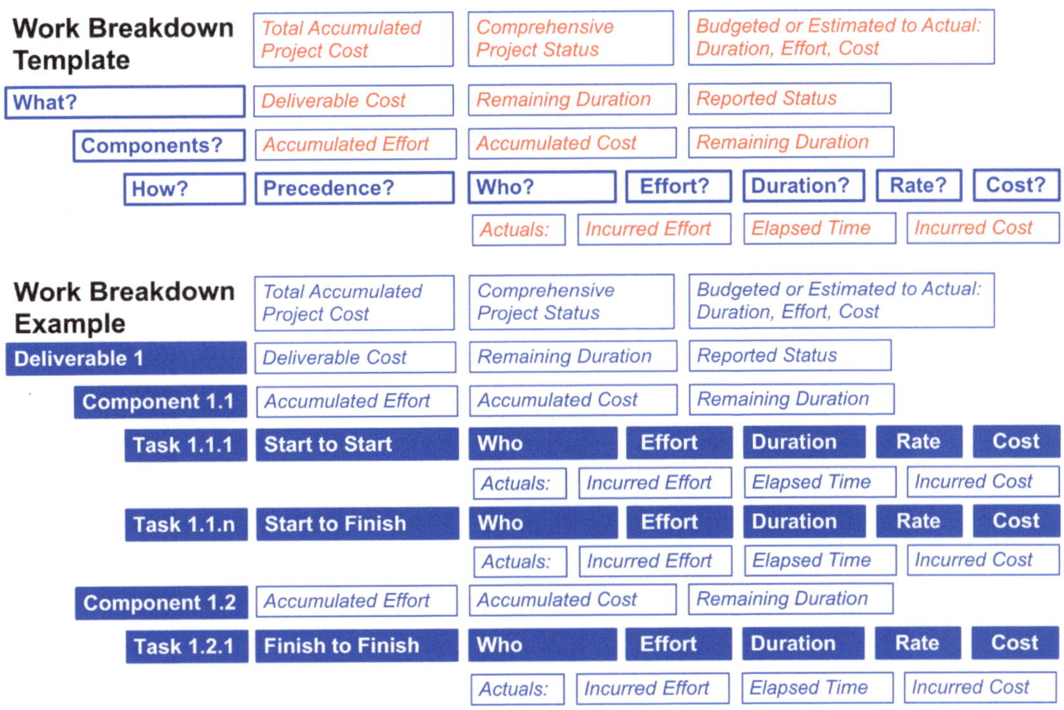

Figure 13-4. Planning Elements

Looking at the work breakdown template, the blue-lettered elements represent how the planning questions assemble into a plan. The red-lettered elements represent the upward accumulation at differing plan levels (task, component, deliverable, and project) of the actual effort, duration, and cost incurred as the plan is worked. The work breakdown example (below the template) represents a portion of a plan with values substituted for questions. Other than coloring, the two structures are the same.

Once a project is under way, determining project status is quite simple – it just needs to be done often and kept current. Checklist 13-9 provides a short list of questions for finding status. Responses to these questions plus completed work paints a good picture of progress (or lack of it). One suggested technique is to ask, "How much is left?", which is the reverse of the more customary approach, "How much has been done?".

Checklist 13-9. Tracking Status

- ❐ Who is having problems, with what, why, and for how long?
- ❐ What tasks have not started, and how long will it be before they start?
- ❐ What deliverables, components, or tasks have not been completed?
- ❐ For each task, how much elapsed time (duration) remains until completion?
- ❐ For each task, how much work (effort) remains to be done until completion?
- ❐ For each task, how much budget remains until completion?
- ❐ Are there any deliverables or components that require (or will require) changes that will affect schedule, effort, and budget in any way (positively or negatively)?

End of Checklist

From responses to the status tracking questions, a project manager can flag positive and negative conditions within the project plan using, the terms and colors below accompanied by explanatory remarks:

- Completed – Green
- In Progress – Blue
- At Risk – Yellow
- Critical – Red
- Temporarily Stopped – Orange
- Discontinued – Black

Finally, "the numbers" are combined with reported status (per the questions) and plan flags to provide a complete "to-date" snapshot of the project in terms of estimated to actual:

1. The variance between planned and projected schedule base on current elapsed time
2. The variance between planned effort and actual work accrued
3. The variance between planned budget and actual accrued costs

Pursuing Quality

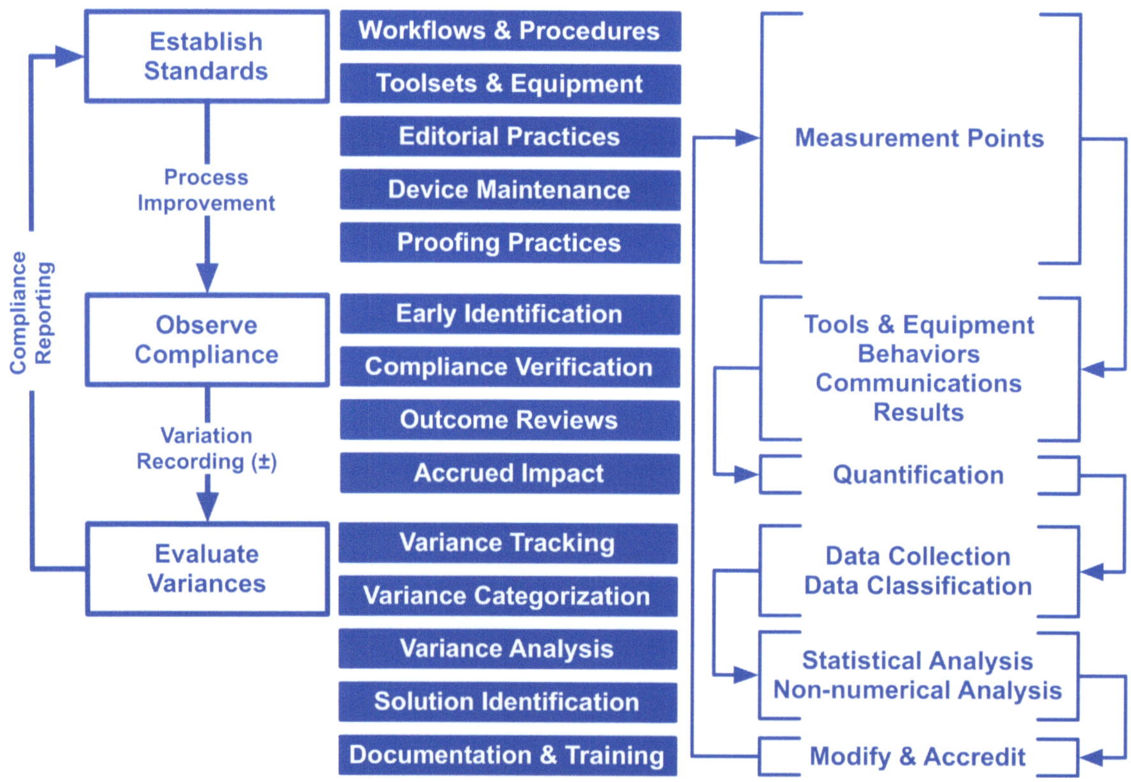

Figure 13-5. Quality: Standards, Observations, & Evaluations

The delivery of consistently high quality products and solutions is dependent upon the removal of variance from the production process. Simple! Yes? Then again, maybe not. Figure 13-5 provides an overview of a closed loop quality system.

We will hit the highlights of standardizing, monitoring for compliance, and making assessments. As usual, our mantra will be, "keep it simple!". The discussion is intended to provide some additional operational knowledge that can be applied to prior chapters and the demonstrations yet to come to improve results.

The material presented in the topic is not all inclusive, nor is it necessarily going to be "on the money" for a given organization. It simply presents a starting point for further thought, investigation, and application. If there is anything needed in variable content endeavors, it is constraining unwanted or unplanned variance – this is one approach.

Standardization

This is an elementary set of standards. It is not comprehensive; it deliberately omits configuration preferences, numeric settings, calibrations, and other typical detail. The aim is to apply elementary operating constraints to minimize potential variance and optimize consistent repeatable production results and products. The approach is to manage variables associated with how work is accomplished and the tools used during the production process to minimize or remove random variance and inconsistency before product output.

- **Standard Workflows & Standard Operating Procedures —** Control or minimize the introduction of variability caused by human error during the performance of assigned activities:
    - Operational procedures
    - Objectives identification
    - Requirements specification
    - Design activities
    - Data and digital content activities
    - Project planning and execution
    - Quality assurance activities
- **Standard Toolset —** Minimize the introduction of variability generated by application software and operating system anomalies. Additionally, ensure that tools and software render consistent predictable results:
    - Approved operating systems
    - Standard application software
    - Standard application software settings
- **Standard Editorial Practices —** Textual content issues may seem relatively simple, although not necessarily true in practice. They can be exceedingly time-consuming when reviewing text and making necessary corrections. The following areas need to be checked and addressed:
    - Composition
    - Accuracy
    - Appropriateness and wording
    - Message content and delivery
    - Spelling and Grammar
    - Punctuation and Capitalization
    - Formatting
    - Correct attribution and reference
    - Potential copyright infringement
- **Device Maintenance, Calibration, & Profiling —** Insure consistent optimal device performance:
    - Source Devices: Digital Cameras, Scanners, Video Recorders, Audio Recorders , and other input devices
    - Bidirectional Devices — Displays, Monitors, and LCDs
    - Destination Devices (proofers, printers, presses, and projection units)

- **Color Proofing Practices**
  Verify major color process milestones for regularity, uniformity, or degree of variance in colors:
  - Origination: Original file measurements
  - Conversion Proof: Work space profile conversion results compared to original file measurements
  - Correction Proof: Corrected file compared to profile conversion measurements
  - Color Conversion and Proof: Color conversion (RGB to CMYK or sRGB) file compared to correction measurements
  - Soft PDF Proof: PDF proof file compared to color conversion measurements
  - Hard Proof: Production PDF proof print compared to conversion measurements and original (unconverted) file measurements

Observation

The critical factor in effective monitoring is achieving balance; the watchword is "No Surprises". The worst possible situations are too much monitoring (big brother) and too little monitoring (dreamland). Here are a few simple ideas to stimulate thinking:

- Early problem identification points
  - Tools and equipment — Usage and Settings
  - Behavior
    - Compliance with standards, procedures, instructions, and specifications
    - Attention to detail
    - Sense of urgency
  - Communicating and documenting information and circumstances
    - Verbal and written communications that are complete, accurate, and timely
    - Documentation that is complete, accurate, and timely
- Standards compliance verification
  - Reviews and walkthroughs
  - Spot checks and Sampling
  - Formal inspections (scheduled and ad hoc)
- Outcome or product characteristics
  - Accuracy and clarity
  - Color tonality and uniformity
  - Conformance with requirements
  - Audio quality and volume
- Problem impact areas
  - Effort expenditure and labor costs
  - Material usage and replacement costs
  - Equipment operation and processing costs
  - Schedule slippage and opportunity costs

Evaluation

To understand the frequency, incidence, and effect of variance in the production process, we need to determine how much documentation from monitoring is required, at what level of detail, and for how long we are going to retain the documentation. Again, there are a variety of approaches with their partisans. As before, what is offered here is food for thought.

- Recorded variances and incidents
  - Tools and equipment
    - Improper use
    - Inaccurate settings or calibration deviation
  - Work Behavior
    - Noncompliance with standards, procedures, instructions, and specifications
    - Inattention to details
    - Insufficient timeliness
  - Communicating and documenting information and circumstances
    - Insufficient, inaccurate, incomplete, or untimely communications
    - Insufficient, inaccurate, incomplete, or untimely documentation
- Categorize and group variances and incidents
  - By year, month, week
  - By organization, project, shift, supervisor, individual
  - By process, procedure, activity, tool, or equipment, job, skill
- Analysis of variances and incidents
  - Mean, median, and standard deviation
  - Frequency and trend analysis
  - Correlation and linear regression
  - Ratio analysis
    - Accumulated variance as a percentage of total project experience
    - Accumulated variance as a percent of compliance
  - Mission completion and risk analysis
  - Accumulated problem impact (time, labor, materials, equipment operation, opportunity costs)
    - Effort
    - Cost
    - Schedule
  - Non-numerical analysis
    - Project and compliance reviews
    - Roundtable discussions
    - Task groups
    - Workshops
    - Process analysis and functional decomposition
- Remedial Planning – identify improvements to standards, procedures, and working practices to enhance workflow and results
  - Brainstorming
  - Roundtable discussions
  - Task groups
  - Workshops
  - Process analysis and solution design

- Document, train, and promote
  - Individual self-directed training
  - Ad hoc reading and instruction
  - Hands-on exercises (individuals and teams)
  - Formal training — In-house sessions or Seminars, classes, and other outside training opportunities
  - Testing and certification
  - Periodic refreshers
  - Performance reviews

Supplementary Reading

There are a variety of publications on project definition and requirements analysis, project management, quality assurance, and related issues that can be viewed from the perspective of digital printing and graphic design. The books suggested below are easily read and instructive. While not specifically about digital print, their perspectives of manufacturing and information technology are quite relevant to our topic.

Eckes, G. (2003). *Six sigma for everyone.* Hoboken, NJ: John Wiley & Sons, Inc.

George, M., Rowlands, D., Kastle, B. (2004). *What is lean six sigma?* New York: McGraw-Hill.

Graham, I. (1998). *Requirements engineering and rapid development: an object-oriented approach.* Harlow, England: Addison Wesley Longman, Limited.

Lewis, J.P. (1991). *Project planning, scheduling & control – a hands-on guide to bringing projects in on time and on budget.* Chicago: Probus Publishing Company.

Paulk, M., Weber, C., Curtis, B., Chrissis, M., et al. (1994). *The capability maturity model: guidelines for improving the software process.* Software Engineering Institute, Carnegie Mellon University. Reading, MA: Addison Wesley Longman, Inc.

Romano, F., et al. (2002). *Designing for digital.* Alexandria, VA: Digital Printing Council, Printing Industries of America.

FIVE: DEMONSTRATIONS

We've talked about it; now let's see it! After exploring a bssic set of tools and techniques for producing variable content documents, this unit disects a series of progressively more involved variable document examples.

14. Demonstration Overview

Chapter 14 is simply a short introduction to the assessment demonstrations. It explains the purpose behind the next five chapters and the approach used in examining them.

a. Background
b. Approach

15. A Simple Text Demonstration

Chapter 15 analyzes the development steps and associated issues in producing a report or an academic document using variable techniques.

a. Purpose
b. Description
c. Components
d. Assembly
e. Observations

16. The "Articles" Demonstration

Chapter 16 looks at development of a newsletter containing a single topic for each weekly issue.

a. Purpose
b. Description
c. Components
d. Assembly
e. Observations

17. The "Editions" Demonstration

Chapter 17 is a continuation of the prior chapter. It combines the weekly topics into a monthly newsletter containing the articles for that month.

a. Purpose
b. Description
c. Components
d. Assembly
e. Observations

18. The "Poems" Demonstration

Chapter 18 looks at the steps involved in creating a poetry collection or anthology composed of the text, related graphic, and bibliographic comments per poem.

a. Purpose
b. Description
c. Components
d. Assembly
e. Observations

19. The "Mailer" Demonstration

Chapter 19 examines creating different versions of a marketing mailer using variable data printing techniques.

a. Purpose
b. Description
c. Components
d. Assembly
e. Observations

20. Cards – Analysis, Conjecture, & Possibility

Chapter 19 examines how variable techniques can be used in making greeting cards. The purpose of the discussion is to highlight document design decisions regarding the use of variable content. It concludes by showing a set of static wedding documents for the reader to redesign or sketch for variable content and features.

a. Purpose
b. Alternatives
c. Foundations
d. Decisions
e. Possibilities

21. A Publishing Encore

Chapter 21 acknowledges the variety of technologies available for producing variable documents. Rather than relying strictly on digital press or high-speed printer technology for production, the focal point of discussion centers on digital desktop technology, such as laser and inkjet (both with duplex print capabilities). Discussion particularly addresses the use of inkjet printing, speciality print stock, and types of ink to produce artistic or customized variable documents in limited run quantities.

a. Implication
b. On Reflection
c. Who & Why?
d. Printing Issues
e. Ink On What?
a. Intimation

14. VCP Demonstration Overview

Discussion Points:

- ✓ **Background**
- ✓ **Approach**

Background

The next six chapters are short demonstrations (show-'n-tell) of documents that were created using combinations of the technologies we've been discussing. The documents we'll be viewing were started as trials to obtain hands-on experience with the tools and techniques we have previously covered.

One aim of the trials was to identify basic tools and possible combinations of steps (inputs, processes, outputs), chain the combinations together in a simple common workflow, and then characterize a nominal skill set required to construct the different documents. Another aim was to gauge the variety of variable documents this simple workflow could produce, while looking for potential complications that may be encountered. The outstanding question, left for another time, is the degree of potential difficulty in training a neophyte to be proficient with the workflow, selected tools, and related techniques.

The documents we'll explore here range from a report to greeting cards, and include newsletters, poetry, and marketing mailers. Obviously, these are not the only documents that can be created, but they present enough variety and information to hopefully stimulate further thinking.

So! These assessments are not explanations about "how" to develop a particular type of document. Rather, they ask questions about document variety and possibility. The implied brainteaser at the end of each assessment is this:

Given document type "D" plus toolkit, techniques, and related skills, what additional document designs "d_1" through "d_n" are possible within that type?

It is left to the reader to determine what might be personally or individually relevant, and it is also up to the reader to craft his or her own reply.

Approach

Each chapter examines a different document. Although potential issues may be similar, they tend to vary in number and complexity in correlation with particular document features: design layout, content type and content size or quantity. Each chapter is organized the same way:

1. Stated purpose – what is being evaluated and why
2. Description of the document
3. Components used to construct the document
4. Considerations in assembling the document
5. Concluding remarks about issues confronted during design and construction, and potential problems that may be encountered in producing similar documents

Regarding conclusions, they are a mix of opinion and hands-on experience. They are certainly not the only conclusions that could be drawn, but are ones deemed noteworthy. It is left to individual readers to form their own opinions. The implication here is that readers may find it worthwhile to try their hand at constructing similar documents using this approach.

The individual demonstrations present differing perspectives on applying the tools and techniques previously discussed, nuances associated with development, and fundamental considerations related to the results obtained in each case.

- Chapter 15: The "Report" Demonstration – an academic report
- Chapter 16: The "Articles" Demonstration – a weekly newsletter
- Chapter 17: The "Editions" Demonstration – a monthly newsletter with multiple stories
- Chapter 18: The "Poems" Demonstration – three poems using a standard layout
- Chapter 19: The "Mailer" Demonstration – two versions of a marketing mailer
- Chapter 20: Cards – Analysis, Conjecture, & Possibility – an exploration of greeting cards

So! Without further delay, let's move on to Chapter 15 and the report....

15. A Simple Text Demonstration

Discussion Points:

- ✓ Purpose
- ✓ Description
- ✓ Components
- ✓ Assembly
- ✓ Observations

Purpose

The purpose of this demonstration is to start with a very simple example of content variability — content in this case being text. The tools and processes we'll be seeing are the same ones we've talked about as individual topics in earlier chapters. Instead of focusing on each technology separately, we'll integrate selected technical capabilities to begin forming a consistent pattern of process approach and tool application.

Description

Content for this demonstration happens to come from an academic research paper. The original was produced with Microsoft Word. It is a formal document, a literature review. For our purposes, what it says and how it looks are irrelevant. It is simply text with a particular hierarchical structure:

- Title Page
  - Title
  - Author
  - Location
  - Date
- Table of Contents
- Introduction
- Methodology
  - Approach
  - Source Selection Criteria
  - Delimitations
- Documents
- The Document Life Cycle
- Focus on Issues
  - Business Processes and Documents
  - Document Management,
  - Workflow Management
  - Document Storage and Retrieval}
- Conclusions
- Bibliography

Components

The XML version of the report has three main components: a document type definition (DTD), a text file tagged with XML as specified by the DTD, and a cascading stylesheet (CSS) that uses the XML tags in formatting the text.

```
<!-- DTD for Report Document -->

<!ELEMENT document (frontsection, contentsection,
      referencesection)>
<!ELEMENT frontsection (titlepage, tableofcontents?)>
<!ELEMENT titlepage (title, subtitle?, author+, course?,
      department?, college?, institute, location,
      publishdate)>
  <!ELEMENT title (#PCDATA)>
  <!ELEMENT subtitle (#PCDATA)>
  <!ELEMENT author (#PCDATA)>
  <!ELEMENT course (#PCDATA)>
  <!ELEMENT department (#PCDATA)>
  <!ELEMENT college (#PCDATA)>
  <!ELEMENT institute (#PCDATA)>
  <!ELEMENT location (#PCDATA)>
  <!ELEMENT publishdate (#PCDATA)>

 <!ELEMENT tableofcontents (contenttitle, content+)>
  <!ELEMENT contenttitle (#PCDATA)>
  <!ELEMENT content (contentitem+, subcontent*)>
   <!ELEMENT contentitem (#PCDATA)>
   <!ELEMENT subcontent (#PCDATA)>

<!ELEMENT contentsection (topic+)>
 <!ELEMENT topic (topictitle, topicpara+, subtopic*)>
  <!ELEMENT topictitle (#PCDATA)>
  <!ELEMENT topicpara (#PCDATA)>
  <!ELEMENT subtopic (subtopictitle, subtopicpara+)>
   <!ELEMENT subtopictitle (#PCDATA)>
   <!ELEMENT subtopicpara (#PCDATA)>

<!ELEMENT referencesection (bibliography?)>
 <!ELEMENT bibliography (bibliotitle, cite+)>
  <!ELEMENT bibliotitle (#PCDATA)>
  <!ELEMENT cite (citeauthor, citedate, citetitle, citesource)>
   <!ELEMENT citeauthor (#PCDATA)>
   <!ELEMENT citedate (#PCDATA)>
   <!ELEMENT citetitle (#PCDATA)>
   <!ELEMENT citesource (#PCDATA)>
```

Listing 15-1. A Report DTD

DTD

Listing 15-1 is the complete Document Type Definition (DTD). This DTD has four purposes:

1. Provide a reference for tagging the document text
2. Validate the XML-tagged document
3. Provide a reference for building the cascading style sheet
4. Provide a standard structure for reports

Notice the similarities between the DTD and the simple report outline on the prior page. Other than conventions expected by the XML parser, the important point with the DTD is the additional detail that our outline just doesn't provide.

XML

Listing 15-2, Listing 15-3, and Listing 15-4 are portions of the report's tagged XML file.

Listing 15-2 is a portion of the front section of the the report – the title page and part of the table of contents (TOC). All are tagged with XML according to the report DTD.

Listing 15-3 is another snippet of the XML file – an extract from the initial content of the report body.

Listing 15-4 provides an example of bibliographic cites that has been tagged with XML per the DTD.

Taken together, this should provide a fair idea of the relationship between the DTD and tagging the report text with XML.

```
<?xml version="1.0" encoding="UTF-8"?>
<?xml-stylesheet type="text/css" media="screen, print" href="Report.css"?>
<!DOCTYPE document SYSTEM "Report.dtd">
<document>
    <frontsection>
        <titlepage>
            <title>The Life Cycle of Business Documents:</title>
            <subtitle>A Literature Review</subtitle>
            <author>Author Name</author>
            <course>Research Methods in Graphic Media</course>
            <department>School of Print Media</department>
            <college>College of Imaging Arts and Sciences</college>
            <institute>Rochester Institute of Technology</institute>
            <location>Rochester, New York</location>
            <publishdate>Month Day, Year</publishdate>
        </titlepage>
        <tableofcontents>
            <contenttitle>Table of Contents</contenttitle>
            <content>
                <contentitem>Introduction</contentitem>
                <contentitem>Methodology</contentitem>
                <subcontent>Approach</subcontent>
                <subcontent>Source Selection Criteria</subcontent>
                <subcontent>Delimitations</subcontent>
            </content>
```

Listing 15-2. XML-tagged Front Matter

```
<contentsection>
    <topic>
        <topictitle>Introduction</topictitle>
        <topicpara>The purpose of this researcher is to survey literature pertaining to the life
            cycle of business documents. Literature dealing only with the document life cycle as
            a whole is scarce. However, literature in a variety of forms containing short
            discussions of the document life cycle is plentiful. This literature includes
            everything from chapters in management books to white papers to one page sales
            pieces.</topicpara>
        <topicpara>The document in one form or another has been around for thousands of years
            (Bergmanns, 1998). Ndlela (2003) in her editorial on document management
            competencies comments that "documents are the natural byproducts of enterprise
            activity. Any information that is created or communicated forms a document."
```

Listing 15-3. Portion of XML from Report Body

```
<referencesection>
    <bibliography>
        <bibliotitle>Bibliography</bibliotitle>
        <cite>
            <citeauthor>Adobe Systems Incorporated.</citeauthor>
            <citedate>(2005).</citedate>
            <citetitle>Adobe creative suite 2 design guide.</citetitle>
            <citesource>San Jose, CA: Adobe Systems Incorporated.</citesource>
        </cite>
```

Listing 15-4. Example XML Bibliographic Citation

CSS

At this point, all the content has been prepared and tested against standards, meaning that it's "well formed" and "valid". Now the issue is formatting the content so that it will be readable and attractive when produced for either print or Web. Let's move on to the cascading stylesheet, Listing 15-5. The stylesheet is based on CSS version 2.1, which provides added presentation capabilities beyond screen display.

As opposed to using strictly HTML tags to apply desired formatting, HTML and XML in combination capitalize on existing tags. Essentially, the XML tags are treated as if they were HTML tags. As a result, document formatting complies with standards defined in the DTD.

However, all the XML tags associated with applying formatting must be defined within the context of CSS to reflect desired format attributes. This is not exactly the WYSIWYG layout approach we've all become accustomed to.

It is also not very compact. You might note that for the sake of brevity, Listing 15-5 only addresses the title page. In its entirety, it contains 160 lines of coding. What we have here is enough for you to see how it reflects the DTD.

As an aside, we're looking at CSS only to demonstrate the power and flexibility of the XML-based approach. In the following chapters, we'll be using a visual tool.

```
/* CSS Stylesheet for Report Document */
/* page setup */
@page normal {
    size: portrait;
    margin: 0.5in;
}
/* document */
* {
    color: black;
    background: white;
    font-family: Arial, Verdana, sans-serif;
    font-size: 12pt;
}
/* frontsection */
/* titlepage */
title {
    font-size: 150%;
    font-weight: bold;
    text-align: center;
    display: block;
    margin-top: 5.0em;
    margin-bottom: 0.5em;
}
subtitle {
    font-size: 140%;
    font-weight: bold;
    text-align: center;
    display: block;
    margin-top: 0.2em;
    margin-bottom: 1.5em;
}
author {
    font-weight: normal;
    text-align: center;
    display: block;
    margin-top: 10.5em;
    margin-bottom: 10.5em;
}
course, department, college, institute, location {
    font-weight: normal;
    text-align: center;
    display: block;
    margin-top: 0.5em;
    margin-bottom: 0.5em;
}
publishdate {
    font-weight: normal;
    text-align: center;
    display: block;
    margin-top: 5.0em;
    margin-bottom: 5.0em;
    page-break-after: always;
}
```

Listing 15-5. Report CSS *Segment*

Assembly

Assembly, if you want to call it that, is very simple – only a matter of joining the external files (DTD and CSS) to the XML text file. When an XML-capable Web Browser processes the XML text file, the browser automatically processes the DTD for validation and the CSS for display formatting. The Web Browser displays the formatted report as one contiguous document. When printed from the Web Browser, the report displays as a paginated document according to CSS instructions.

This first instruction after the XML declaration (reference Listing 15-2) links the XML document and the stylesheet. The instruction denotes the stylesheet type as CSS text. It defines the display media as both screen and print, and provides the location and file name of the stylesheet. Here is the extracted instruction from the XML file that makes the link:

```
<?xml-stylesheet type="text/css" media="screen, print"
    href="Report.css"?>
```

The next XML file instruction after the CSS link (see Listing 15-2) simply links the XML document to its validating DTD file, "Report.dtd". Here is the DTD link:

```
<!DOCTYPE document SYSTEM "Report.dtd">
```

Observations

This is one way to produce a standardized document layout that can be used to produce additional reports with the same formatting but different content about different topics. Additional formatting effort is not required – it's all dependent on the XML tags. Formatting the XML report as demonstrated provides a fairly satisfactory rendition of the original document. Follow-on reports can be written and tagged with a simple text editor. A word processor is not necessary. The documents are browser-ready and reasonably printable. However, consider this:

First, browser-supplied information is added as a footer. It includes the display date and the XML file URI. On one hand, it could be an unneeded distraction in the printed version. On the other, it does provide additional source documentation that could be useful later.

Second, the table of contents is useless as is. Different solutions depend on how the document will primarily be displayed: display only, print only, or both. Here are three suggestions:

1. Redesign the DTD, stylesheet, and XML tagging to accommodate existing page numbering. The assumption is that pagination generated by browser and word processor will be the same. Perhaps, but don't count on it!
2. For a print-only version, change the tagging of the table of contents to include the page numbers. Selecting this option may require fine tuning.
3. For a display-only version, hyperlink the table of contents entries to the appropriate headings in the text. When viewing the document on-screen, the reader can navigate to desired sections of the report. Forget the page numbering, but be sure to include back links to allow the reader to return to the table of contents.

One more issue needs mentioning. Advocates of XSL-FO (Extensible Stylesheet Language Formatting Objects) may consider replacing CSS as an alternative approach to putting XML in print. Replacing CSS with XSL-FO may add additional complexity to the process, although the end results may be cleaner as far as print goes. However, possible lack of a visual development environment may add other complications. In terms of developer acceptance, visual development is the primary and preferred avenue for expressing creativity – not issuing textual instructions in the hope of quickly achieving accurate results.

In contrast, as far as visual development for CSS, a product such as Adobe Dreamweaver already exists. It is fully featured, and compatible with various versions of HTML and XML. In fact, a number of visual tools and approaches for CSS are available. For those who may want to further explore CSS capabilities, the two references below may be informative. These are just two from a wide array of information sources that might be helpful:

Meyer, Eric A. (2007). CSS: the definitive guide, third edition. Sebastopol, CA: O'Reilly Media, Inc.

Schmitt, Christopher. (2004). CSS cookbook. Sebastopol, CA: O'Reilly Media, Inc.

16. The "Articles" Demonstration

Discussion Points:

- ✓ Purpose
- ✓ Description
- ✓ Components
- ✓ Assembly
- ✓ Observations

Purpose

In the previous chapter, we looked at establishing a simple XML-based CSS framework to produce additional reports with new content while adhering to structural and formatting standards. In light of observations about this approach, let's examine how to obtain similar variable capabilities using visual tools to produce more satisfying results.

Description

Our document is a flier with a weekly article about some experience related to gardening or nature. Each article is one or two pages long and includes a relevant photograph. Figure 16-1 is an example of this document.

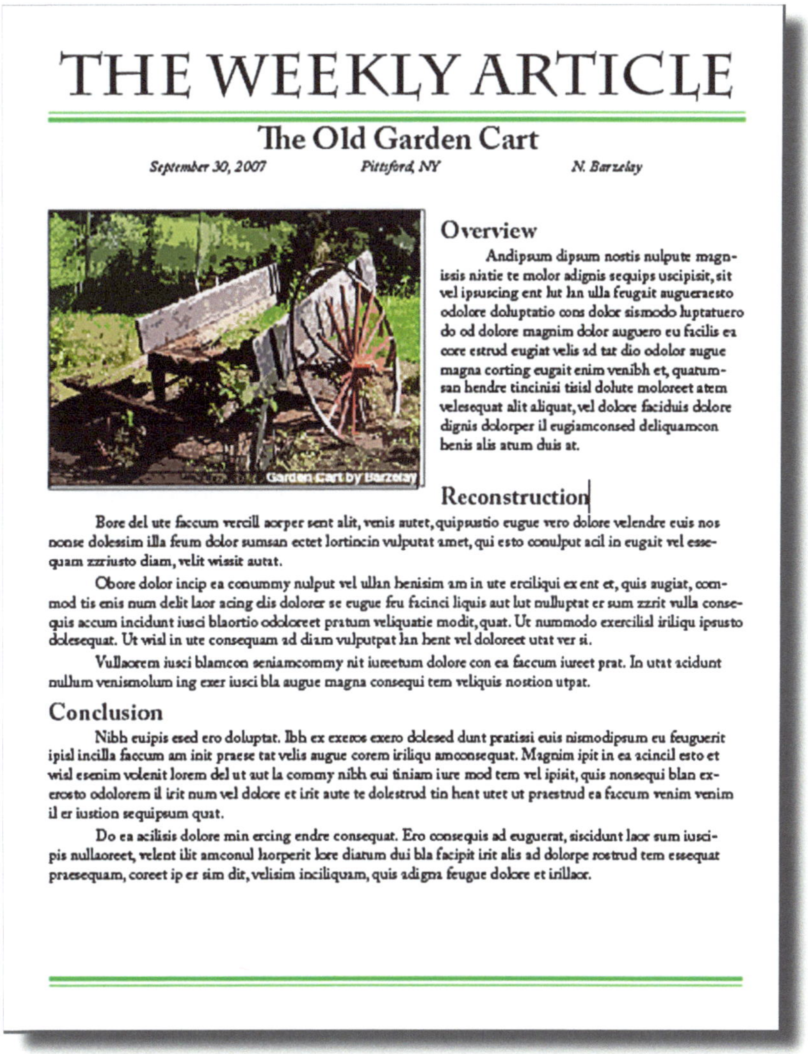
THE WEEKLY ARTICLE

The Old Garden Cart

September 30, 2007 *Pittsford, NY* *N. Barzelay*

Overview

Andipsum dipsum nostis nulpute magnissis niatie te molor adignis sequips uscipisit, sit vel ipsuscing ent lut lan ulla feugait augueraesto odolore doluptatio cons dolor sismodo luptatuero do od dolore magnim dolor auguero eu facilis ea core estrud eugiat velis ad tat dio odolor augue magna corting eugait enim venibh et, quatumsan hendre tincinisi tisisl dolute moloreet atem velesequat alit aliquat, vel dolore faciduis dolore dignis dolorper il eugiamconsed deliquamcon benis alis atum duis at.

Reconstruction

Bore del ute faccum vercill aorper sent alit, venis autet, quipsustio eugue vero dolore velendre euis nos nonse dolessim illa feum dolor sumsan ectet lortincin vulputat amet, qui esto conulput acil in eugait vel essequam zzriusto diam, velit wissit autat.

Obore dolor incip ea conummy nulput vel ullan henisim am in ute erciliqui ex ent et, quis augiat, commod tis enis num delit laor acing elis dolorer se eugue feu facinci liquis aut lut nulluptat er sum zzrit vulla consequis accum incidunt iusci blaortio odoloreet pratum veliquatie modit, quat. Ut nummodo exercilisl iriliqu ipsusto dolesequat. Ut wisl in ute consequam ad diam vulputpat lan hent vel doloreet utat ver si.

Vullaorem iusci blamcon seniamcommy nit iureetum dolore con ea faccum iureet prat. In utat acidunt nullum venismolum ing exer iusci bla augue magna consequi tem veliquis nostion utpat.

Conclusion

Nibh euipis esed ero doluptat. Ibh ex exeros exero dolesed dunt pratissi euis nismodipsum eu feuguerit ipisl incilla faccum am init praese tat velis augue corem iriliqu amconsequat. Magnim ipit in ea acincil esto et wisl esenim volenit lorem del ut aut la commy nibh eui tiniam iure mod tem vel ipisit, quis nonsequi blan exerosto odolorem il irit num vel dolore et irit aute te dolestrud tin hent utet ut praestrud ea faccum venim venim il er iustion sequipsum quat.

Do ea acilisis dolore min ercing endre consequat. Ero consequis ad euguerat, siscidunt laor sum iuscipis nullaoreet, velent ilit amconul horperit lore diatum dui bla facipit irit alis ad dolorpe rostrud tem essequat praesequam, coreet ip er sim dit, velisim inciliquam, quis adigna feugue dolore et irillaor.

Figure 16-1. The Weekly Article – An Example

Components

Document construction is based on same the principle of standardization employed in the previous chapter. There is a small collection of components that include prepared content and structural elements:

- Document type definition (DTD) file
- XML-tagged text file
- Digital image file
- Document layout file

DTD

The DTD (Listing 16-1) defines a standard structure for all variable content used in each article.

```
<?xml version="1.0" encoding="UTF-8"?>

<!-- DTD for Newsletter Article -->

<!ELEMENT article (title, dateline, image, caption, text)>

<!ELEMENT title (#PCDATA)>

<!ELEMENT dateline (date, location, author)>
 <!ELEMENT date (#PCDATA)>
 <!ELEMENT location (#PCDATA)>
 <!ELEMENT author (#PCDATA)>

<!ELEMENT image EMPTY>
    <!ATTLIST image href CDATA #REQUIRED>

<!ELEMENT caption (#PCDATA)>

<!ELEMENT text (subtitle,para+)*>
 <!ELEMENT subtitle (#PCDATA)>
 <!ELEMENT para (#PCDATA)>
```

Listing 16-1. Weekly Article DTD

XML

Using the DTD, the variable content is encoded with XML tags (Listing 16-2 on the next page). That XML-tagged content includes a reference to the photograph used in the particular article. That reference gives the location and name of the photographic image (its URI). Shown here is the usual syntax for such references:

```
<image href="file://folderName/filename.tiff"/>
```

An example of usage is the yellow highlighted line of code in the XML listing below.

```xml
<?xml version="1.0" encoding="UTF-8"?>
<!DOCTYPE article SYSTEM "Article.dtd">
<article>
<title>The Old Garden Cart</title>

<dateline>
<date>September 30, 2007</date>
<location>Pittsford, NY</location>
<author>N. Barzelay</author>
</dateline>

<image href="file://GardenCart.jpg"/>
<caption>Garden Cart by Barzelay</caption>

<text>
<subtitle>Overview</subtitle>
<para>Andipsum dipsum nostis nulpute magnissis niatie te molor adignis sequips uscipisit, sit vel ipsuscing ent lut lan
    ulla feugait augueraesto odolore doluptatio cons dolor sismodo luptatuero do od dolore magnim dolor auguero
    eu facilis ea core estrud eugiat velis ad tat dio odolor augue magna corting eugait enim venibh et, quatumsan
    hendre tincinisi tisisl dolute moloreet atem velesequat alit aliquat, vel dolore faciduis dolore dignis dolorper il
    eugiamconsed deliquamcon henis alis atum duis at.</para>
<subtitle>Reconstruction</subtitle>
<para>Bore del ute faccum vercill aorper sent alit, venis autet, quipsustio eugue vero dolore velendre euis nos nonse
    dolessim illa feum dolor sumsan ectet lortincin vulputat amet, qui esto conulput acil in eugait vel essequam zzri-
    usto diam, velit wissit autat.</para>
<para>Obore dolor incip ea conummy nulput vel ullan henisim am in ute erciliqui ex ent et, quis augiat, commod tis
    enis num delit laor acing elis dolorer se eugue feu facinci liquis aut lut nulluptat er sum zzrit vulla consequis accum
    incidunt iusci blaortio odoloreet pratum veliquatie modit, quat. Ut nummodo exercilisl iriliqu ipsusto dolesequat.
    Ut wisl in ute consequam ad diam vulputpat lan hent vel doloreet utat ver si.</para>
<para>Vullaorem iusci blamcon seniamcommy nit iureetum dolore con ea faccum iureet prat. In utat acidunt nullum
    venismolum ing exer iusci bla augue magna consequi tem veliquis nostion utpat.</para>
<subtitle>Conclusion</subtitle>
<para>Nibh euipis esed ero doluptat. Ibh ex exeros exero dolesed dunt pratissi euis nismodipsum eu feuguerit ipisl
    incilla faccum am init praese tat velis augue corem iriliqu amconsequat. Magnim ipit in ea acincil esto et wisl
    esenim volenit lorem del ut aut la commy nibh eui tiniam iure mod tem vel ipisit, quis nonsequi blan exerosto
    odolorem il irit num vel dolore et irit aute te dolestrud tin hent utet ut praestrud ea faccum venim venim il er
    iustion sequipsum quat.</para>
<para>Do ea acilisis dolore min ercing endre consequat. Ero consequis ad euguerat, siscidunt laor sum iuscipis nul-
    laoreet, velent ilit amconul laorperit lore diatum dui bla facipit irit alis ad dolorpe rostrud tem essequat praese-
    quam, coreet ip er sim dit, velisim inciliquam, quis adigna feugue dolore et irillaor.</para>
</text>
</article>
```

Listing 16-2. Weekly Article XML

Layout Design

That leaves the layout for the document. Rather than CSS, the layout was created and formatted using Adobe InDesign. A common (and tempting) approach in preparing the document layout for loading text is shown in Figure 16-2. The layout is designed with a box for each XML-tagged content element. This may work for a single static document, but for variable content documents, it has distinct disadvantages. Why? Because the document layout has been "tailor-fitted" to a specific body of content.

Just for example, say that a particular document has a one paragraph introduction, four paragraphs in the body, and three paragraphs in the conclusion.

What happens when the next article has a different number of paragraphs for the introduction, body, and conclusion? What if the next article is less than a page or requires two pages?

What happens is a lot more work, plain and simple. The so-called standardized layout is not standard after all – it is customized for each iteration of the document. It will require rework for each new article.

All this can be avoided by making sure that all repeating static content is expressed in the document page masters, and the area that will receive the XML content (both text and images) is defined as a single master text frame as in Figure 16-3 (next page).

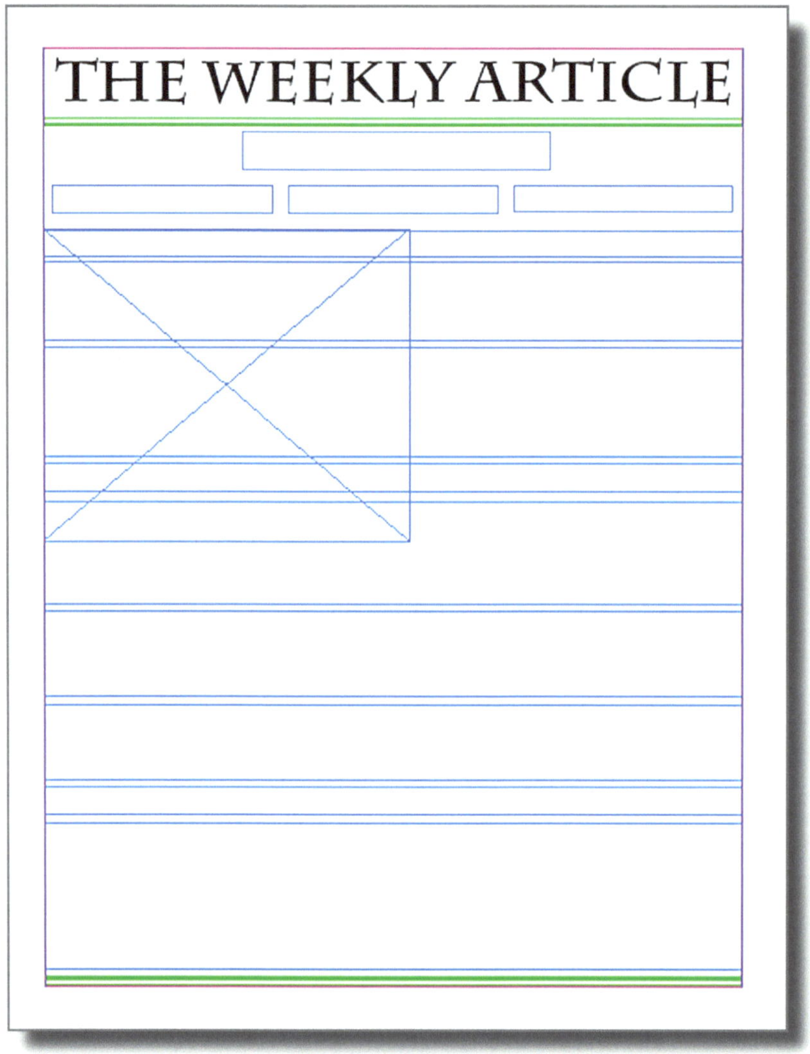

Figure 16-2. A Common Design Approach

All placement and formatting issues are addressed using paragraph, character, and object styles mapped to XML tags. No matter how the text differs in length, number of headings, paragraphs, even images, results will meet expectations without structural design rework.

As you can see, there's not much to revise. The Adobe InDesign processor uses the XML and the related styles to automatically flow and format the text into the document.

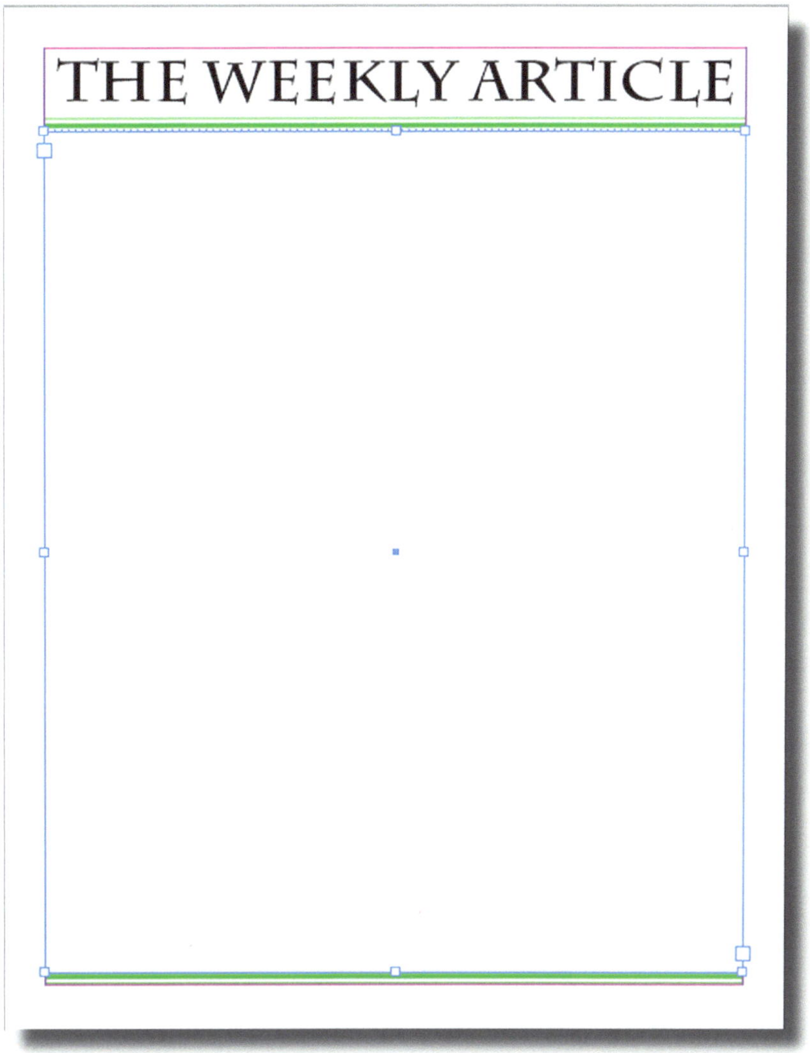

Figure 16-3. Weekly Article Designed With a Single Master Text Frame

Assembly

After setting up the first set of components, it is only a matter of loading the XML into the the document layout. (See Chapter 10, "Document Design -- Integration & Print Stream" for details.) Producing new articles only requires tagging text, referencing any digital content (imagery) in appropriate XML tags according to the DTD, and then importing the new material into the InDesign document layout.

Figures 16-4 and 16-5 (next page) are examples – both produced from the same DTD and design layout used for Figure 16-1 at the beginning of this chapter. Please feel free to compare the three documents. You will see that design standards have been met. At the same time, new content is accommodated with each document version.

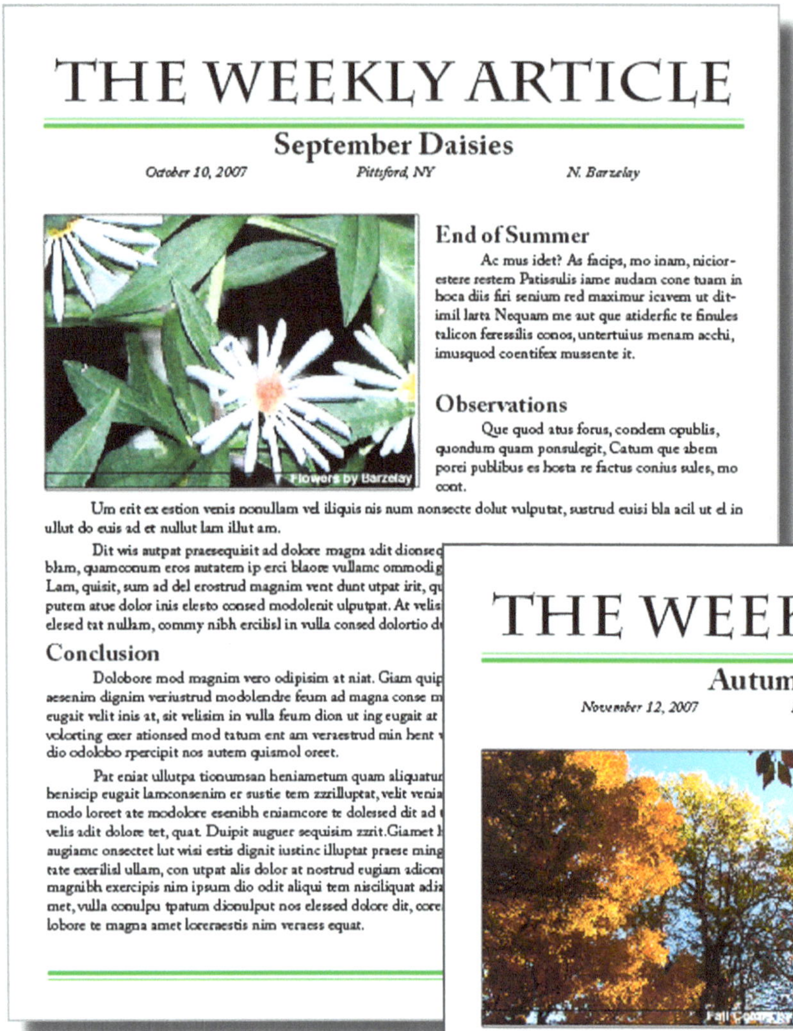

Figure 16-4. Weekly Article XML Reload

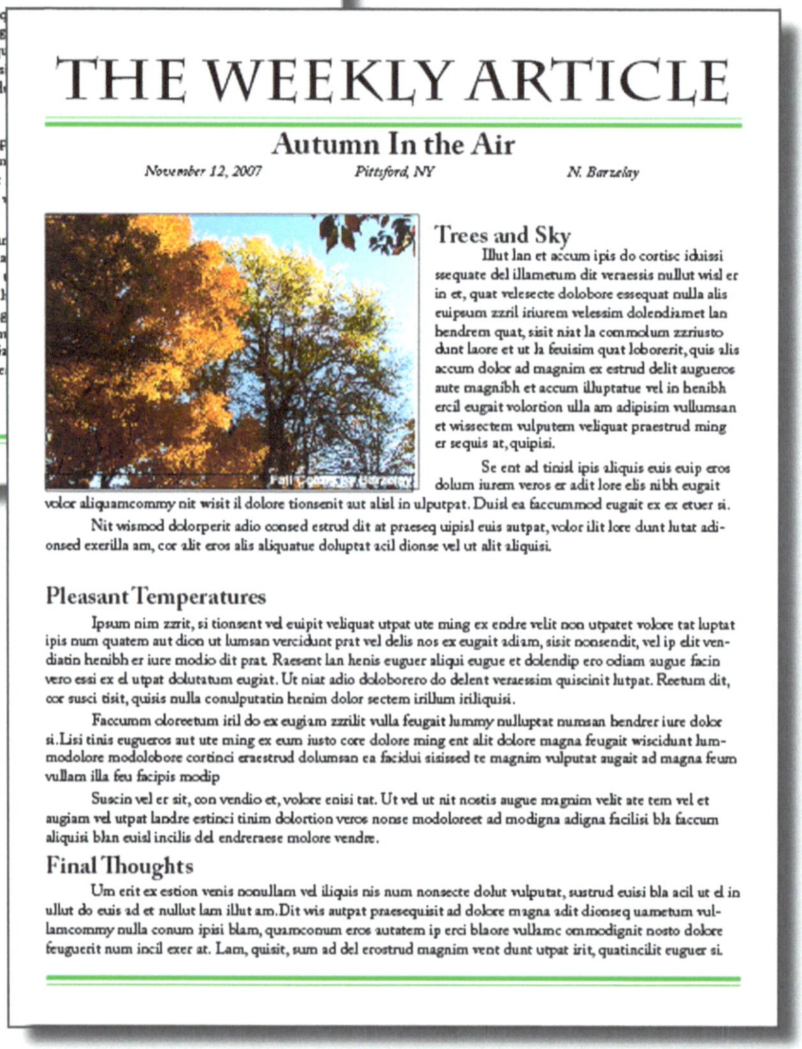

Figure 16-5. Another XML Reload

Observations

Unlike the printed document in the previous chapter, our printed output no longer has the potentially problematic text inserted by the Web Browser at the end of each page. Using the visual document layout tools and approach in this chapter's example, it would be fairly easy to redo the academic report. The only change would be the replacement of the cascading stylesheet with a document layout file that would include paragraph and object styles mapped to the XML.

The downside of this revised approach is that it does not provide the simple dual media capability by itself. An easy approach for going to the Web would be the development of a CSS document linked into the XML. A more complex approach would be the use of InDesign export facilities and Adobe Dreamweaver to set up the Web capability.

In terms of looking beyond this document, it is easy to imagine expanding on this theme. Albeit a "baby step", that is what we are going to do in the next chapter.

17. The "Editions" Demonstration

Discussion Points:

- ✓ Purpose
- ✓ Description
- ✓ Components
- ✓ Assembly
- ✓ Observations

Purpose

In the last chapter we looked at processing single documents where formatting was based upon standard XML tags and mapped styles imbedded in the document layout file. Content for succeeding versions of a particular subject – a new report or article – could be loaded into this predefined layout. We were dealing with a "one-at-a-time" manually fed process, although formatting was automated in some measure. Once the XML content was imported, the new document could be reproduced many times.

Our intention is to move from processing a single document for a single article to laying out a document that contains a collection of individual articles within a single document . Each element in the collection could potentially be an individual document in its own right. We want to group them into an integrated whole. The end result will still be a single document containing the same content and reproduced multiple times.

Description

The new document is an edition of a monthly newsletter, "The Marsh Monthly". It represents a collection of articles similar to those covered in the prior chapter. In fact, the "monthly edition" is very much like the "weekly article", except that it packages a variable number of articles together. It loads both text and images for a varying number (1 to n) of variable length articles. It automates additional page generation as needed to accommodate the collection. With this in mind, let's survey the various components needed to assemble the collection into an integrated document.

Components

The components that make up this new document are much the same as what we have already seen. This list of items should look very familiar:

- A DTD
- An XML-tagged text stream
- Front and trailing page layout masters
- A set of digital images (one for each article in the collection)

```
<?xml version="1.0" encoding="UTF-8"?>

<!-- DTD for Document Edition -->

<!ELEMENT edition (masthead?, article*)>

<!ELEMENT masthead (eddate, edno, edpub, edloc)>
 <!ELEMENT eddate (#PCDATA)>
 <!ELEMENT edno (#PCDATA)>
 <!ELEMENT edpub (#PCDATA)>
 <!ELEMENT edloc (#PCDATA)>

<!ELEMENT article (title, dateline, image, caption, text)*>

<!ELEMENT title (#PCDATA)>

<!ELEMENT dateline (date, location, author)>
 <!ELEMENT date (#PCDATA)>
 <!ELEMENT location (#PCDATA)>
 <!ELEMENT author (#PCDATA)>

<!ELEMENT image EMPTY>
   <!ATTLIST image href CDATA #REQUIRED>

<!ELEMENT caption (#PCDATA)>

<!ELEMENT text (subtitle?,para+)*>
 <!ELEMENT subtitle (#PCDATA)>
 <!ELEMENT para (#PCDATA)>
```

Listing 17-1. DTD for Monthly Newsletter

DTD

Other than the masthead and the need for two page layouts (the first page and any succeeding pages), the main difference between this DTD (Listing 17-1) and the one from the prior chapter is that we have to account for a collection of multiple articles and images.

The indicator in the DTD to validate one or more articles is set in the "edition" element by the stared subelement "article*". Refer to the yellow highlighted element in Listing 17-1.

XML

Notice that the XML presented in Listing 17-2 (next page) only includes the first of the three articles contained in the full XML content stream. The other two will be much the same – at least in the fact that they also will validate against the same DTD in Listing 17-1.

We have a couple of options for loading the content. We could load the three XML text files one at a time in the sequence we want the articles to appear. Another alternative would be to combine the files into a single longer file in the desired sequence. The larger file will still validate with the DTD. We would then load the entire multi-article file at one time. Either approach is fine – whichever seems easier for you.

```
<?xml version="1.0" encoding="UTF-8"?>
<!DOCTYPE edition SYSTEM "Edition.dtd">

<edition>
<masthead>
<eddate>October</eddate>
<edno>no.1</edno>
<edpub>NDB</edpub>
<edloc>Rochester, NY</edloc>
</masthead>

<article>
<title>Into The Blue</title>
<dateline>
<date>September 30, 2007</date>
<location>Pittsford, NY</location>
<author>N. Barzelay</author>
</dateline>
<image href="file://Flight.jpg"/>
<caption>Flight by Barzelay</caption>
<text>
<subtitle>History</subtitle>
<para>Faccumm oloreetum iril do ex eugiam zzrilit vulla feugait lummy nulluptat numsan hendrer iure dolor si.</para>
<para>Lisi tinis eugueros aut ute ming ex eum iusto core dolore ming ent alit dolore magna feugait wiscidunt lum-
    modolore modolobore cortinci eraestrud dolumsan ea facidui sisissed te magnim vulputat augait ad magna feum
    vullam illa feu facipis modip</para>
<subtitle>Impressions</subtitle>
<para>Suscin vel er sit, con vendio et, volore enisi tat. Ut vel ut nit nostis augue magnim velit ate tem vel et augiam vel
    utpat landre estinci tinim dolortion veros nonse modoloreet ad modigna adigna facilisi bla faccum aliquisi blan
    euisl incilis del endreraese molore vendre.</para>
<para>Um erit ex estion venis nonullam vel iliquis nis num nonsecte dolut vulputat, sustrud euisi bla acil ut el in ullut do
    euis ad et nullut lam illut am.</para>
<para>Dit wis autpat praesequisit ad dolore magna adit dionseq uametum vullamcommy nulla conum ipisi blam,
    quamconum eros autatem ip erci blaore vullamc ommodignit nosto dolore feuguerit num incil exer at. Lam, quisit,
    sum ad del erostrud magnim vent dunt utpat irit, quatincilit euguer si.</para>
<subtitle>Conclusion</subtitle>
<para>Lortion ullaortisi tet alit ulputem atue dolor inis elesto consed modolenit ulputpat. At velisi tet ipsusto dolobore
    tat. Ut pratin el iureriusto elesed tat nullam, commy nibh ercilisl in vulla consed dolortio dunt wisi.</para>
</text>
</article>

<article>
<title>Faithful Friend</title>
<dateline>

etcetera, etcetera...

</article>
</edition>
```

Listing 17-2. Monthly Newsletter XML Content (1 of 3 similar articles)

Design Layout

Figure 17-1 is the layout for this multi-page document. Once again, it is very simple. Visually, it only contains the static images and text that you see, plus master text frames – one for each column. In front is the layout for the first page. Behind is the layout for any subsequent document pages. Essentially, the two page masters are the same in terms of processing.

Both page masters use the same styles mapped with the XML tags. Once styles and XML tags are mapped together, they are applied page after page in the initial document and are available for use with any new documents where replicating the same "look" is desirable.

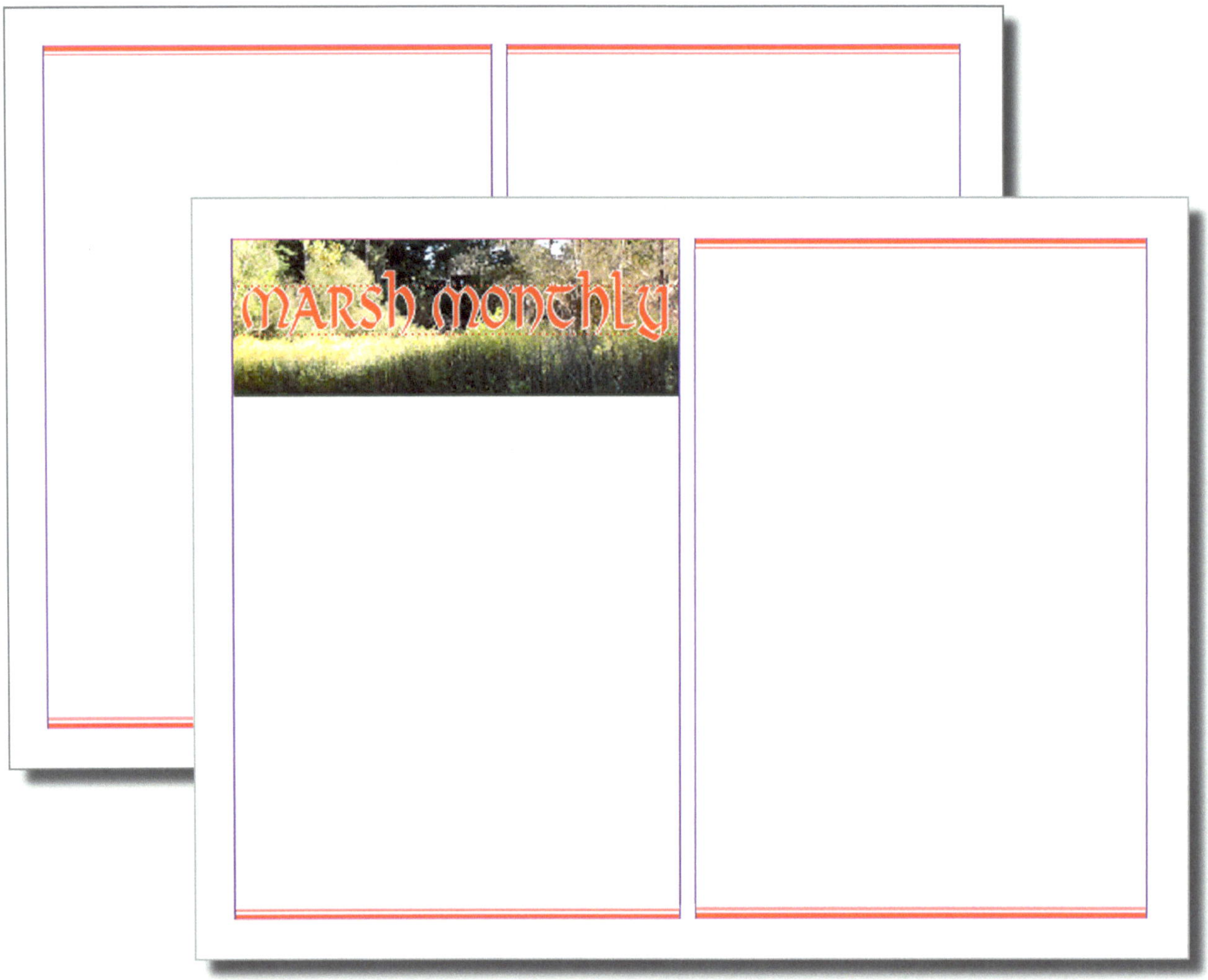

Figure 17-1. Page Masters for First Page (front) Subsequent Pages (back)

Assembly

After all the components have been individually created, import the XML content stream into the document layout, loading a page spread from the front page master first. Then drag a succeeding page master as the next page spread and initiate the automated page generation and loading process from there with a "shift-click" of the mouse. (See Chapter 10, "Document Design -- Integration & Print Stream" for details.)

Columns are linked so that XML flows from the bottom of the left column to the top of the right. For each subsequent pages, XML flows from the bottom corner, right column of the prior page to the top corner, left column of the new page. XML continues to flowcontinues to flow until it has been completely placed into the document. Figure 17-2 is an example of a single edition of the "Marsh Monthly", containing three articles.

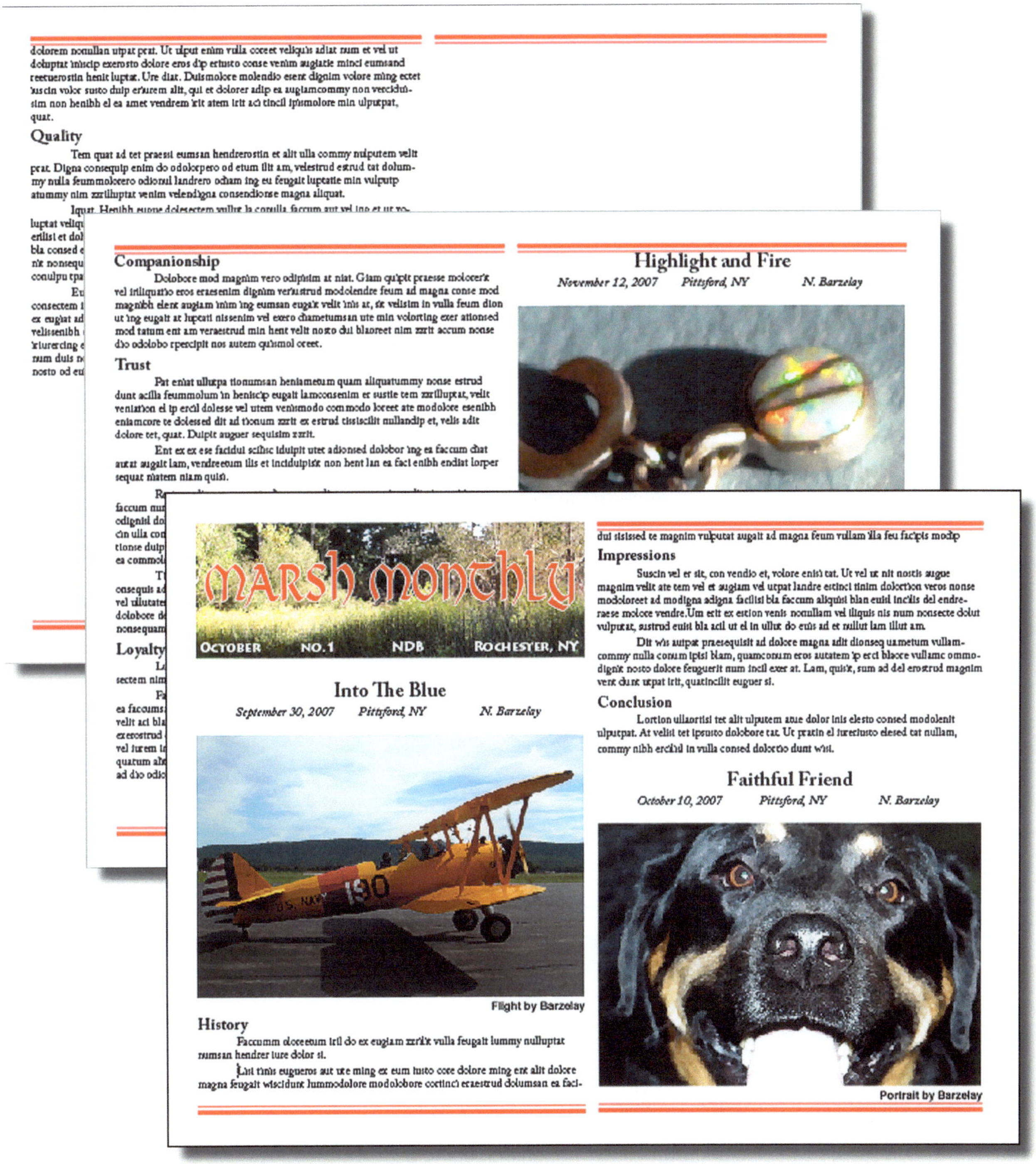

Figure 17-2. Monthly Newsletter (Pages 1 to 3)

Observations

My primary comment is that the current design does not take into account all the variable possibilities that present themselves:

1. The image and text in the masthead could be variable – for instance, the image could vary according to season or according to month.
2. While there is considerable flexibility in the structure of the text – subject headings and number of paragraphs – in each piece, images are limited to one per article. The idea of more images or no images for a single article could be more appropriate.
3. Print space and pagination might be reconsidered for more space economy and improved design integration.

As a matter of note, because the design and production process is based on XML, document content structure and organization is hierarchical. We started out with a single document based on a single subject. Now we have a document that is a collection of subjects. It would not be difficult to expand the hierarchy by classifying the subject matter within our collection to produce a document composed of related topical sections containing sets of subjects. Another possibility is a collection of the year's monthly editions into a year-end annual.

In other words, in terms of variability and extensibility, the processes we've been examining are quite open-ended. We'll continue with this thought in the next chapter.

18. The "Poems" Demonstration

Discussion Points:

- ✓ Purpose
- ✓ Description
- ✓ Components
- ✓ Assembly
- ✓ Observations

Purpose

Now that we've had some exposure to this development process, let's move to a slightly different problem -- a collection or anthology. We'll only deal with three entries, though more are possible

Description

Our subject is a very small set of poems (three, to be exact). These could be the seeds of a much larger collection. Two poems are in English; the third is in French. Each poem contains a little biographical information about the author and a digital photograph meant to provide a little imagery.

For simplicity and to avoid distraction, each poem has its own page – these aren't long poems. That isn't to imply that longer poems extending over multiple pages couldn't be supported with this approach. But poem length and pagination are not of interest as far as this demonstration goes.

Our interest is about dealing with variability. Obviously, one poem is different from the next in terms of meter, rhyme, subject, and general structure. By "structure" I mean organization and presentation – number of stanzas, number of lines within each stanza, and any resulting variability from differing line counts between stanzas or differing numbers of stanzas between poems. The selections demonstrate variation within a given poem and between poems. Accommodating variance facilitates automation.

Components

In terms of components, nothing much has changed. Again, we've got all the usual suspects:

1. A DTD that defines an XML structure suitable for all the poems (Listing 18-1 below)

```
<?xml version="1.0" encoding="UTF-8"?>
<!-- DTD for Poems -->

<!ELEMENT collection (poem+)>
<!ELEMENT poem (title, image?, author?, biography?,
    permission?, stanza+) >

<!ELEMENT title (#PCDATA)>
<!ELEMENT image EMPTY>
  <!ATTLIST image href CDATA #REQUIRED>
<!ELEMENT author (#PCDATA)>
<!ELEMENT biography (#PCDATA)>
<!ELEMENT permission (#PCDATA)>
<!ELEMENT stanza (#PCDATA)>
```

Listing 18-1. The Working Poem DTD

2. An XML text stream based on the DTD and containing the three tagged poems (Listing 18-2 just to the right). Note that only one poem is shown in the XML, but I'm sure you get the picture

3. A set of digital images – one for each poem

4. A page master layout and accompanying styles coordinated with the XML tags (Figure 18-1 on the next page). The layout master simply contains a master text frame – nothing more.

```
<?xml version="1.0" encoding="UTF-8"?>
<!DOCTYPE Poetry SYSTEM "PoetryRev.dtd">

<collection>
<poem>
<title>Chanson D'Automne</title>
<image href="file://Autumn.jpg"/>
<author>Paul Verlaine</author>
<biography>March 30, 1844 to January 8,
    1896</biography>
<permission>In the public domain</permission>
<stanza>
Les sanglots longs
Des violons
De l'automne
Blessent mon coeur
D'une langueur
Monotone.
</stanza>
<stanza>
Tout suffocant
Et blême, quand
Sonne l'heure,
Je me souviens
Des jours anciens
Et je pleure;
</stanza>
<stanza>
Et je m'en vais
Au vent mauvais
Qui m'emporte
Deçà, delà,
Pareil à la
Feuille morte.
</stanza>
</poem>
<poem>

etcetera, etcetera...

</poem>
</collection>
```

Listing 18-2. An XML-tagged Poem

By the way, you might note that the poem in Listing 18-2 is in French. The point is, XML tags can be placed around text in any language. The tags themselves could also be in an other language. There's no reason the XML tags couldn't be French.

And to enhance your reading enjoyment — a little rest break, so to speak — you'll find just below an English translation by a French literary scholar should you happen to not speak French.

Autumn Song

With long sobs
The violin-throbs
Of autumn wound
My heart with languorous
And monotonous
Sound.

Choking and pale
When I mind the tale
The hours keep,
My memory strays
Down other days
And I weep;

And I let me go
Where ill winds blow,
Now here, now there,
Harried and sped,
Even as a dead
Leaf, anywhere.

Figure 18-1. The Poems' Page Master

Assembly

We have seen the assembly process before. It's just like those in previous assessments. There is really nothing new here, except that in setting up the XML load, we need to account for the fact that we are dealing with a high degree of rather random variability throughout the text. The random variability in the text adds an extra factor in setting up the DTD (explained later) and the XML load. If existing XML will be appended, there is no additional concern. However, if the XML will be overwritten, then an additional setting needs to be selected in the XML import setup.

In preparing to load the XML for processing, we'll make some choices in a dialog sequence. First, select the XML file name for import, the "show import options" choice, and the merge option. Then continue to the next dialog box. The following choices need to be selected:

- Clone repeating text elements
- Import only elements that match existing structures
- Don't import contents of "white space only" elements
- Delete elements, frames, and content that do not match imported XML.

This last setting is key in handling the degree of variability in the XML.

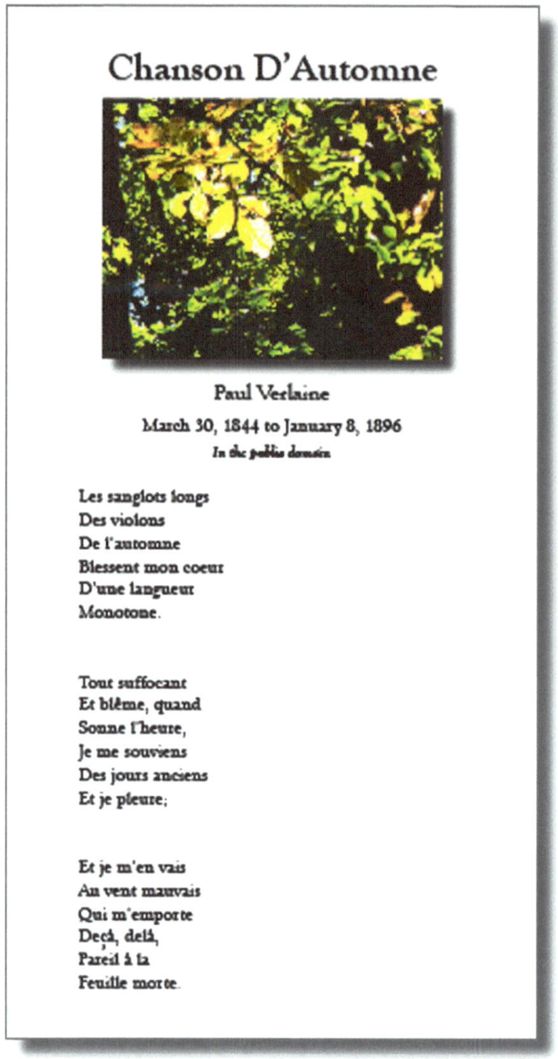

Chanson D'Automne

Paul Verlaine

March 30, 1844 to January 8, 1896

In the public domain

Les sanglots longs
Des violons
De l'automne
Blessent mon coeur
D'une langueur
Monotone.

Tout suffocant
Et blême, quand
Sonne l'heure,
Je me souviens
Des jours anciens
Et je pleure;

Et je m'en vais
Au vent mauvais
Qui m'emporte
Deçà, delà,
Pareil à la
Feuille morte.

Figure 18-2. Example Poem One

The variability issue is more evident when you view the fully assembled collection of poetry. Note the variations within the stanzas; I'm referring to the number of lines, not line length, word count, or character count.

Look closely at Figures 18-2 on this page. Then turn the page and examine Figures 18-3 and 18-4. You may want to flip back and forth, but I'm sure you'll understand. Line counts may easily vary beteween different poems. But note too, line counts withing stanzas in the same poem may also vary, as in Figure 18-4.

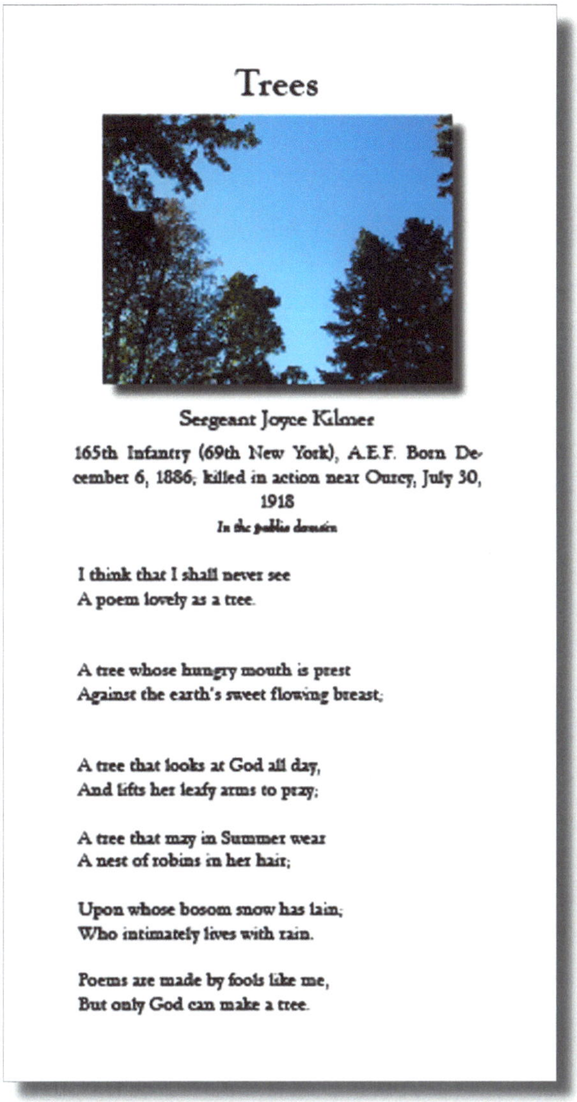

Trees

Sergeant Joyce Kilmer

165th Infantry (69th New York), A.E.F. Born December 6, 1886; killed in action near Ourcy, July 30, 1918

In the public domain

I think that I shall never see
A poem lovely as a tree.

A tree whose hungry mouth is prest
Against the earth's sweet flowing breast;

A tree that looks at God all day,
And lifts her leafy arms to pray;

A tree that may in Summer wear
A nest of robins in her hair;

Upon whose bosom snow has lain;
Who intimately lives with rain.

Poems are made by fools like me,
But only God can make a tree.

Figure 18-3. Example Poem Two

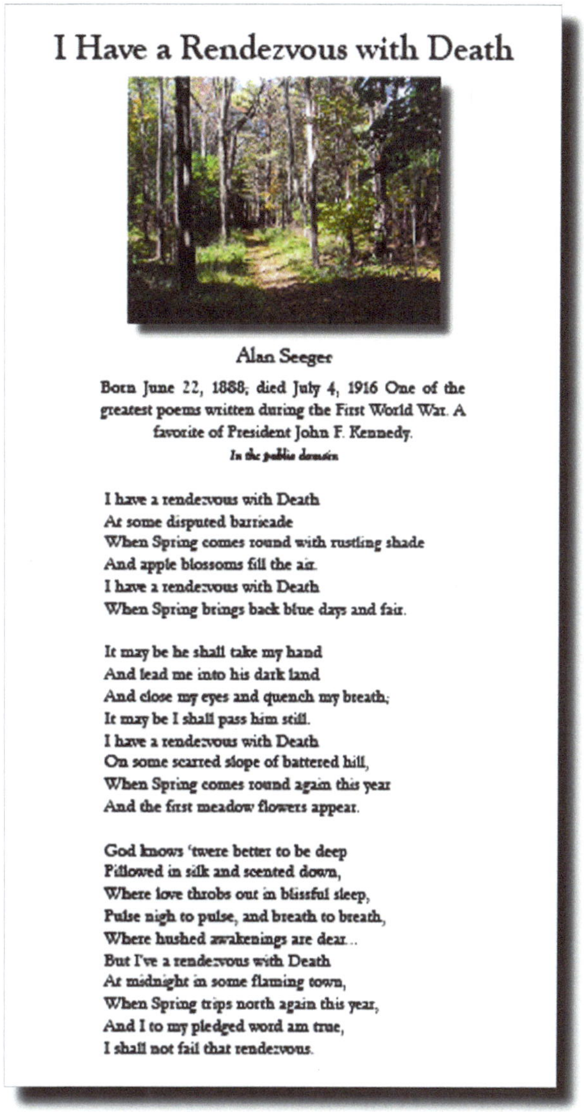

I Have a Rendezvous with Death

Alan Seeger

Born June 22, 1888; died July 4, 1916 One of the greatest poems written during the First World War. A favorite of President John F. Kennedy.

In the public domain

I have a rendezvous with Death
At some disputed barricade
When Spring comes round with rustling shade
And apple blossoms fill the air.
I have a rendezvous with Death
When Spring brings back blue days and fair.

It may be he shall take my hand
And lead me into his dark land
And close my eyes and quench my breath;
It may be I shall pass him still.
I have a rendezvous with Death
On some scarred slope of battered hill,
When Spring comes round again this year
And the first meadow flowers appear.

God knows 'twere better to be deep
Pillowed in silk and scented down,
Where love throbs out in blissful sleep,
Pulse nigh to pulse, and breath to breath,
Where hushed awakenings are dear...
But I've a rendezvous with Death
At midnight in some flaming town,
When Spring trips north again this year,
And I to my pledged word am true,
I shall not fail that rendezvous.

Figure 18-4. Example Poem Three

Observations

Addressing the random variability of the content is an issue that starts at the front end of the development process – defining the content structure in the DTD. Words that come to mind are accuracy, thoroughness, adaptability, and of course the phrase, "attention to detail".

However, there may be limits to how detailed the DTD has to be. With the kind of content we are dealing with in this chapter, too much detail may be dysfunctional. The DTD in Listing 18-4 (below) fully takes into account all common structural elements down to the line level (highlighted in yellow). This was the initial DTD used in the trial. It didn't work for the whole group of poems and had to be modified. Compare this DTD with the successful one shown earlier in Listing 18-1 on page 152.

```
<?xml version="1.0" encoding="UTF-8"?>

<!-- DTD for Poems -->


<!ELEMENT collection (poem+)>

<!ELEMENT poem (title, image?, author?, biography?, permission?, stanza+) >

<!ELEMENT title (#PCDATA)>

<!ELEMENT image EMPTY>
   <!ATTLIST image href CDATA #REQUIRED>

<!ELEMENT author (#PCDATA)>
<!ELEMENT biography (#PCDATA)>
<!ELEMENT permission (#PCDATA)>
<!ELEMENT stanza (line+)>

<!ELEMENT line (#PCDATA)>
```

Listing 18-3. Initial Poem DTD – Line Detail Impedes Needed Flexibility

Now look back at the three poems and pay special attention to lines and stanzas. Note that Listing 18-1 only goes to the stanza level. Why? Because we are able to process the differences in variability at the stanza level. We are not able to obtain satisfactory results at the line level. Ok, so?

It is very similar to adding fractions. To add dissimilar fractions together, we have to find the least common denominator. To process the poems as a group, we have to find the smallest common structural "denominator" – the stanza. We constrain variability to a manageable level by removing potentially infinite variance in grouped material.

Lines create too much variability, so we must go one level higher. At the stanza level we can get the results we want.

19. The "Mailer" Demonstration

Discussion Points:

- ✓ Purpose
- ✓ Description
- ✓ Components
- ✓ Assembly
- ✓ Observations

Purpose

Till now demonstrations focused on text – content that's variable in substance, structure, and size. Now we'll assess an advertising mailer where the source of variable content is a database or other structured digital file of repeating data – equivalent columns within unique rows. We're going to examine "planned" variability rather than random variability.

Description

We'll look at two versions of a marketing mailer. Both are duplex documents, and both use the same source data. The first mailer (Figure 19-1) has one variable side and one static side. Elements on the variable side of the document change with each succeeding record (data row). The variable elements are the addressee's first name, an image, store location and discount imbedded within repeating (or cloned) text, a mailing address and barcode. The other side is entirely static text.

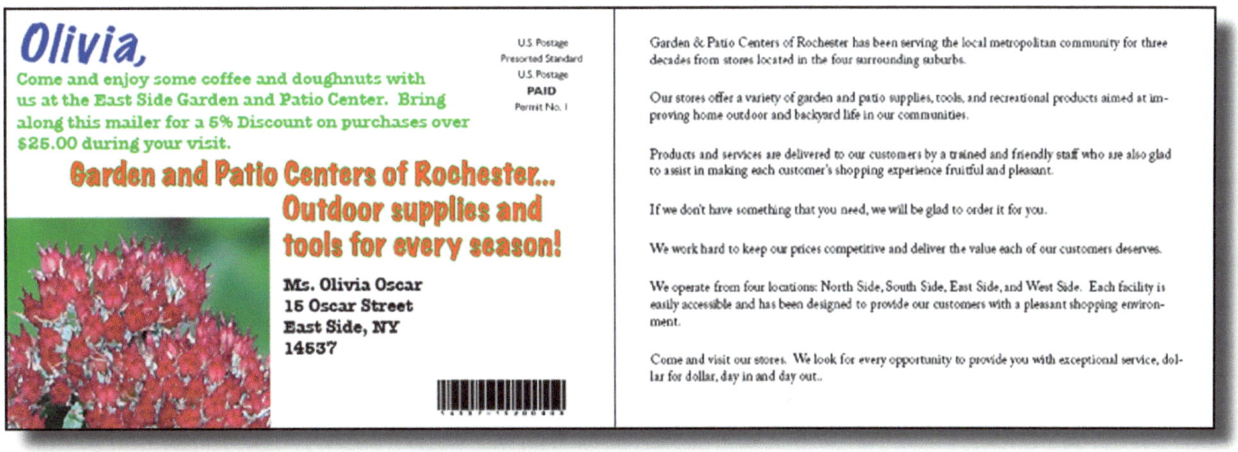

Figure 19-1. Advertising Mailer (Variable on One Side)

In the second mailer (Figure 19-2) both document sides contain variable data. The variable data on the the first side consists of the addressee's first name, an image, and store location, discount, product name and message, all imbedded in the cloned text. Most of the second page is static text except for the data-driven mailing address and barcode.

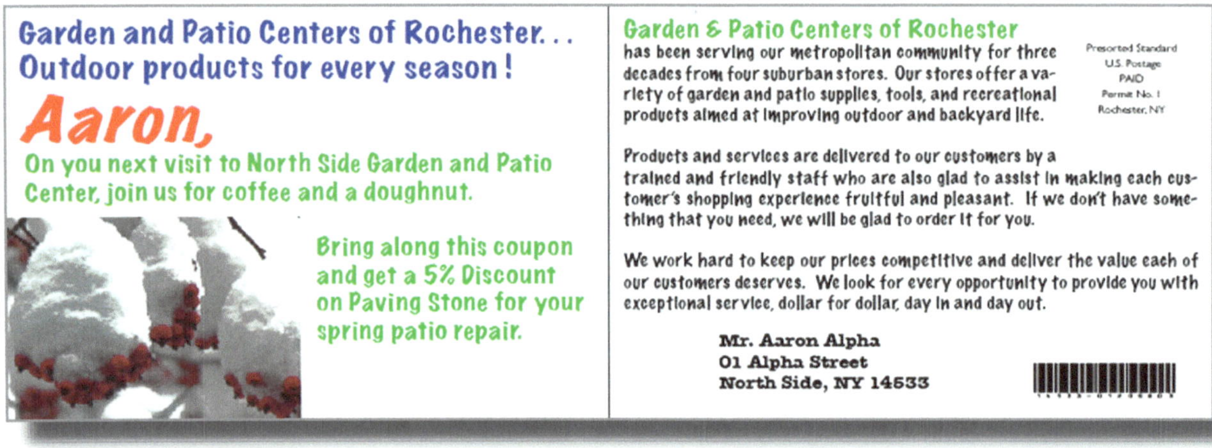

Figure 19-2. Advertising Mailer (Variable Front and Back)

Components

You can probably guess the components by now: DTD, XML data stream, digital images, and document layout. The DTD (Listing 19-1) reflects the structure of the data source – essentially rows and columns (records and fields).

The dataroot (highlighted in yellow) provides for one or more rows of XML content. The row definition immediately follows in the next element.

The XML data stream (Listing 19-2 next page) adheres to the DTD, which in turn reflects the source data structure. The XML data stream shown in Figure 19-4 contains only two records. It has been extracted for demonstration purposes from a much larger set of records (52, if you're curious).

```
<?xml version="1.0" encoding="UTF-8"?>
<!-- DTD for Mailer -->
<!ELEMENT dataroot (row+)>
<!ELEMENT row
     (name,image,location,offer,product,title,first,last,street,city,state,zip,code) >
<!ELEMENT name (#PCDATA)>
<!ELEMENT image EMPTY>
<!ELEMENT location (#PCDATA)>
<!ELEMENT offer (#PCDATA)>
<!ELEMENT product (#PCDATA)>
<!ELEMENT title (#PCDATA)>
<!ELEMENT first (#PCDATA)>
<!ELEMENT last (#PCDATA)>
<!ELEMENT street (#PCDATA)>
<!ELEMENT city (#PCDATA)>
<!ELEMENT state (#PCDATA)>
<!ELEMENT zip (#PCDATA)>
<!ELEMENT code (#PCDATA)>
<!ATTLIST image href CDATA #REQUIRED>
```

Listing 19-1. Advertising Mailer DTD

```
<?xml version="1.0" encoding="UTF-8"?>
<!DOCTYPE dataroot SYSTEM "Mailer.dtd">
<dataroot>
<row>
<name>Anthony</name>
<image href="file://Berries.jpg"/>
<location>North Side</location>
<offer>5% Discount</offer>
<product>Paving Stone</product>
<title>Mr.</title>
<first>Anthony</first>
<last>Able</last>
<street>27 Able Street</street>
<city>North Side</city>
<state>NY</state>
<zip>14533</zip>
<code>14533-27200803</code>
</row>
<row>
<name>Rhonda</name>
<image href="file://Iris.jpg"/>
<location>South Side</location>
<offer>10% Discount</offer>
<product>Fertilizer</product>
<title>Mrs.</title>
<first>Rhonda</first>
<last>Romeo</last>
<street>18 Romeo Street</street>
<city>South Side</city>
<state>NY</state>
<zip>14535</zip>
<code>14535-18200803</code>
</row>
</dataroot>
```

Listing 19-2. Advertising Mailer XML

Small differences between the two mailer versions required some minor adjustments in the DTD and XML to accommodate the different marketing offers presented in each. Otherwise, the rest of the data is the same.

The document layouts, however, are quite different. Figure 19-3 is the layout for the first mailer (Figure 19-1) with one variable and one completely static side.

Figure 19-4 (next page) is the layout for the second mailer (Figure 19-2) which contains variable data on both document sides.

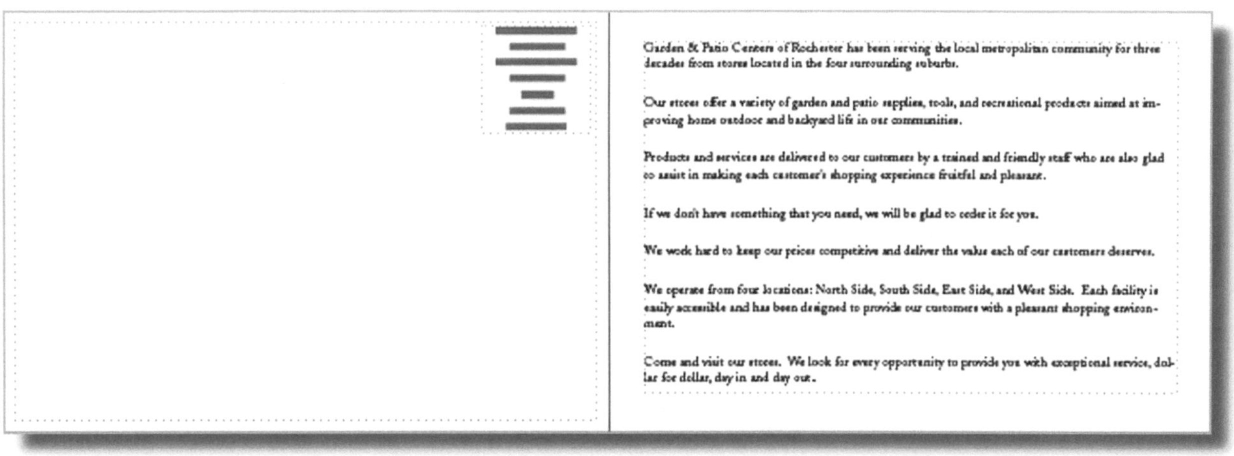

Figure 19-3. Mailer Page Master – Variable on One Side

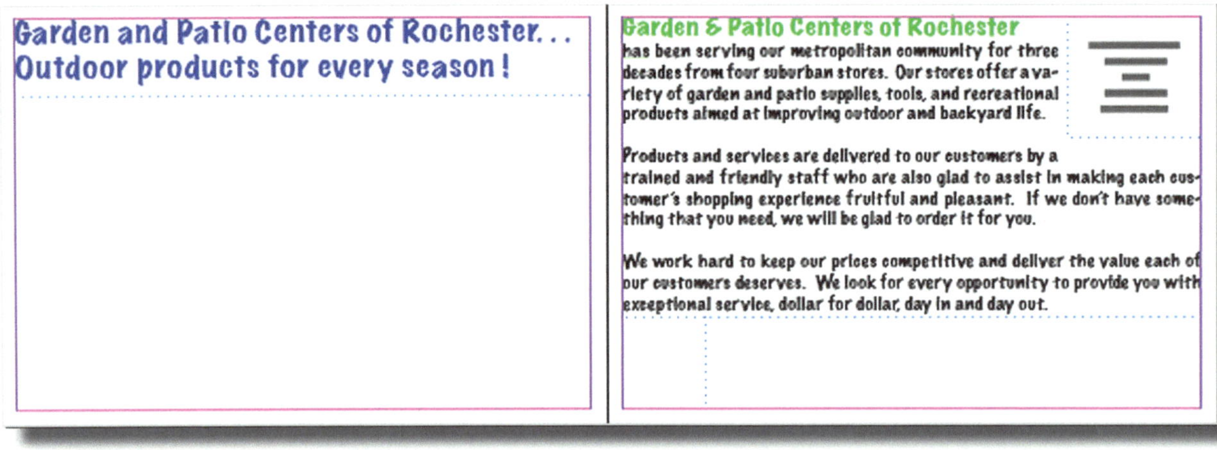

Figure 19-4. Mailer Page Master – Variable Front and Back

Assembly

The process of integrating all the document components is the same as in previous chapters. It's during integration of this type of document that the process really shines. By loading the first document as a starting point, we can kick off an automated process that can correctly generate hundreds or a few thousand documents without any additional human intervention. Figure 19-5, below, and Figures 19-6 and 19-7 on the next page, are print samples generated during content stream processing for the second mailer (Figure 19-2). Document generation was almost instantaneous – for that few documents, it had better be!

Before moving to my observations, take a few moments to examine these mailers and note their variances. You'll see that it's more than just names , images, and addresses.

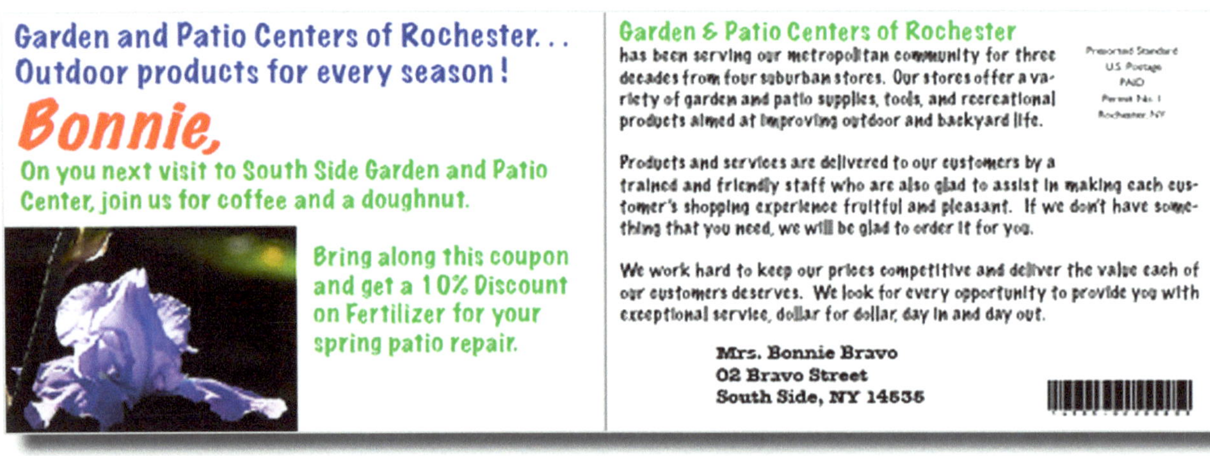

Figure 19-5. Mailer Sample 1

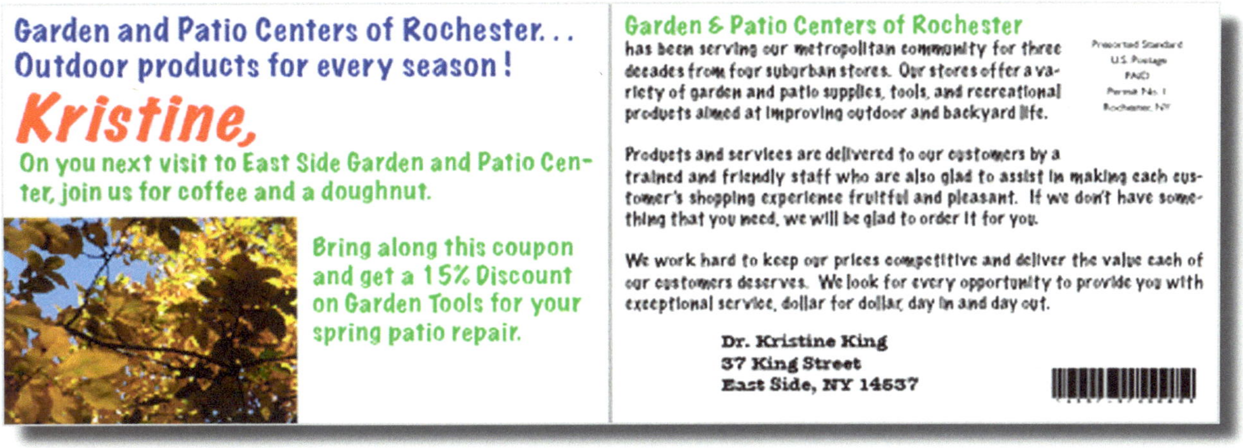

Figure 19-6. Mailer Sample 2

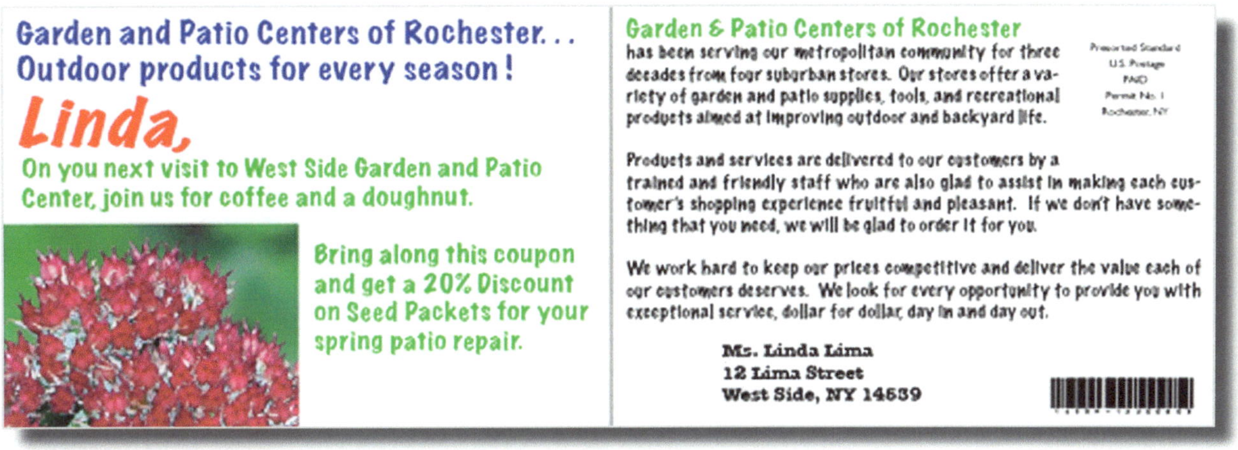

Figure 19-7. Mailer Sample 3

Observations

So what – more variable examples! What's new here that could be interesting? We've looked at the process, the components, variations in text, variations with images, nuances regarding the DTD and XML. What else is there?

To fully appreciate the marketing mailer example – particularly Figures 19-2, 19-5, 19-6, and 19-7 – let's put this discussion in a business context to talk about why this illustration is important. Though this example demonstrates yet another side of variable publishing and printing, it also demonstrates its application to business.

The business in this example, as you can gather, is a garden center. It has four locations. The company tracks business activity at the store level, and consolidates upward. The business activity of interest for this marketing campaign is store inventory, customer transaction trends (what they predominately purchase and where they predominately shop), along with customer response to this particular kind of marketing effort.

The business runs an ongoing discount program tied into their check verification and credit system, so considerable data is collected about each customer at the store level. Customer activity is tied into the individual stores' accounting systems (which include inventory control). Therefore, a lot of business data is available for analysis and decisions. Assuming the use of all this available data, let's look at the different variable elements and the reasoning behind them:

- **Greeting** – the decision was made to send the offer to the predominant shopper or primary card holder. The business maintains good customer relations and felt comfortable using the customer's first name.
- **Store location** – this is the store where the customer primarily shops, although they may visit other store locations too.
- **Discount** – this has to do with the particular store, and is based on the volume of business and the profitability of the store. The store with the lowest business volume merited the highest discount. The idea was to increase activity with a more attractive offer. The store with the highest volume offered the lowest discount.
- **Product** – the decision on the product to include in the offer was based upon a balance between average customer buying preferences, inventory overstock, and inventory activity or movement at each store.
- **Barcode** – this is composed of the customer zip code, customer number, and campaign start date. The purpose is to better understand from which areas each store predominately draws, campaign response rate, and how much additional inventory activity the campaign generated beyond the offer.
- **Image** – in the example, the image was selected by caprice. However, the image could be based on season, gender, store, product offered, or any other kind of criteria.
- **Mailing address** – need we discuss this, except to say that the campaign went to customers of record.

The campaign was not a generalized mailing to "occupant", for example. That is not to imply that another campaign with similar objectives couldn't be run as a generalized campaign to non-customers. There are a thousand and one combinations and possibilities. The only caution would be to avoid sending offers to known, and possibly exceptional customers, with a greeting of "occupant", "resident", "to our friends at...", or some other nondescript verbiage (garbage). After a business has had a relationship with a customer over a period of time, there's no reason to be so impersonal.

The one item missing from these mailer examples is a postal barcode. Adding one is no big deal – primarily adding an additional XML element to the DTD, and adding a postal barcode style to the design layout. A zip code is already available in the address, so it's just a matter of copying and inserting it for each row. Appendix B provides Perl program templates that could be used to automate the process. No need to go back to the data source. No need to do it manually.

What this example shows is the capability of this approach to put VCP in the hands of local or small businesses without large technology investments. In terms of managing aspects of the business and managing the customer relationship, this application has capabilities that extend well beyond the functionality of mail-merge. This is a business solution – not a product.

20. Cards – Analysis & Conjecture

Discussion Points:

- ✓ Purpose
- ✓ Alternatives
- ✓ Foundations
- ✓ Decisions
- ✓ Possibilities

Purpose

We've spent a lot of time examining an assortment of issues related to the development process. What we've seen is that the components and processes in development are quite consistent irregardless of content, composition, complexity, and quantity. Now let's switch our attention from development to formulation — a conceptual mixture of analysis, conjecture, form, and function.

Alternatives

Consider cards – greeting cards, not playing cards (potentially interesting as well). Why the card idea? We'll use greeting cards as our example because they are not – or don't have to be – complex documents. They can be rather simple and yet effective. That also means we can keep our discussion simple and not bog down in innumerable details. Here's the story:

December 25th, Christmas, was coming quickly and cards in stores were too "politically correct", inane, saccharine, suggestive, or just plain vulgar. The alternatives were either to make one or go without. The question: how do you make a card on an old desktop inkjet that only prints on one side? Figure 20-1 shows the answer.

This card was produced with Adobe Illustrator and Photoshop. It has no variable content. However, that doesn't mean it couldn't. It simply becomes a matter of examining possibilities making choices.

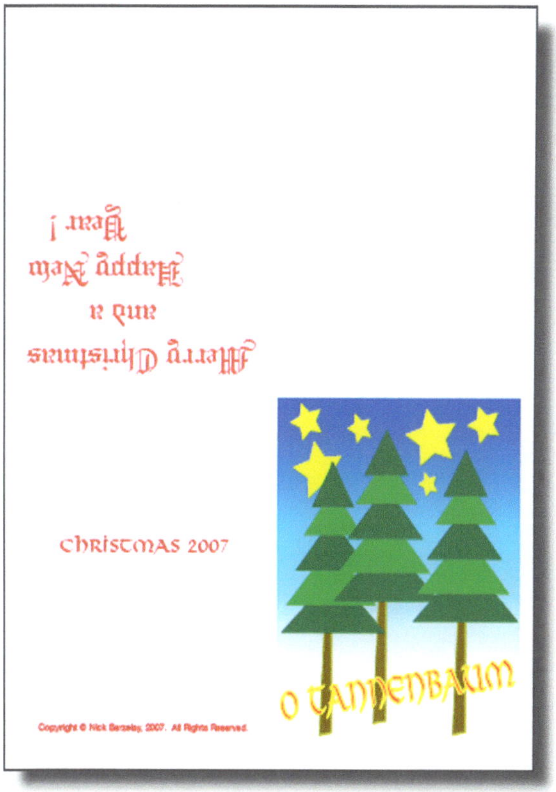

Figure 20-1. Greeting Card Mockup #1

Foundations

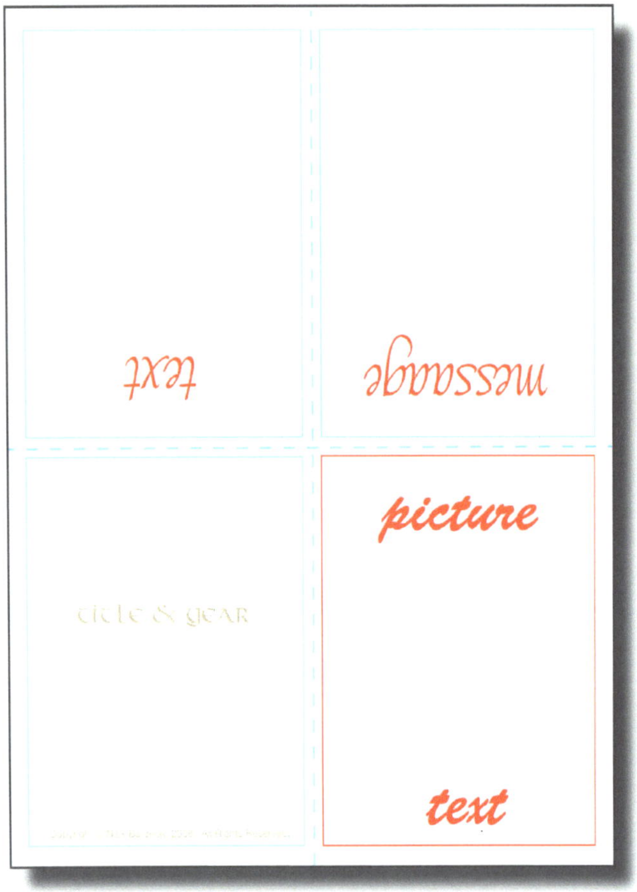

Figure 20-2. Greeting Card Template

The first step to converting from a static card to a variable one is to move to a toolset that will support variable content. While the digital imagery could still originate with Adobe Illustrator or Photoshop, document creation needs a more procedural environment. The obvious choice, for me anyway, is Adobe InDesign.

Figure 20-2 is a depiction of the card layout master in Adobe InDesign. Any of the placeholder text may be made variable or remain static.

The facilitating mechanism for what will be variable is the XML data stream. This can be constructed from a database table, spreadsheet, delimited file, or another XML-tagged text file.

As explained in an earlier chapter, the database provides the most flexibility and control, followed by the spreadsheet.

Figure 20-3 shows such a table – or at least a portion of one. Most of the data fields in this particular table focus on the recipient. We know that recipients will vary, but decisions on other variable content have not been made.

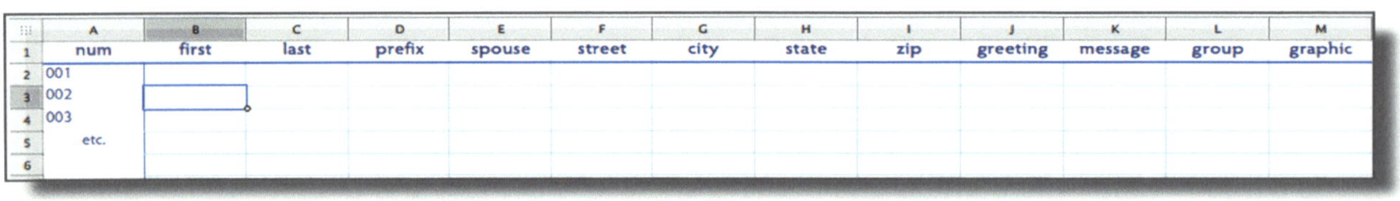

	A	B	C	D	E	F	G	H	I	J	K	L	M
1	num	first	last	prefix	spouse	street	city	state	zip	greeting	message	group	graphic
2	001												
3	002												
4	003												
5	etc.												
6													

Figure 20-3. Greeting Card Data Source

Decisions

There is a point where conjecture, analysis, and toying with possibilities has to end. Decisions are needed! The first decision concerns document type – what it is going to be. In this case, we have already answered that question. The document of interest is a greeting card. All the remaining decisions stem from this single choice. These decisions are fairly standard and consistently applicable to producing almost any document (including both static and variable varieties). While not the only questions to resolve over the course of a project, the ones itemized below need answers to start work (you've seen these questions or similar ones before):

1. What is the document's purpose; what is it intended to accomplish?
2. Who will receive the document and how will it be used?
3. What content in the document will be static?
4. What content in the document will be variable?
5. What data, text, and digital imagery is required for the document?
6. What will be the dimensions of the document?
7. What printing capabilities will be available or required to produce the document?
8. What type and weight of print stock will be needed to produce the document?
9. What other materials and supplies will be needed for the project?

Of course, some choices preclude others. Some aims are mutually exclusive; others seem to be although they're not. A few contradictions may occasionally pop up, and there are always tradeoffs to consider.

For example, let's view a second Christmas card mock-up (Figure 20-4). This card follows the same formula as the earlier one (Figure 20-1). Essentially, the layout is the same, but the content – image, text, and fonts – has changed.

With the intention of using the InDesign master spread presented earlier (Figure 20-2), we'll work our way through a set of nine startup questions on the next page and look at some plausible responses:

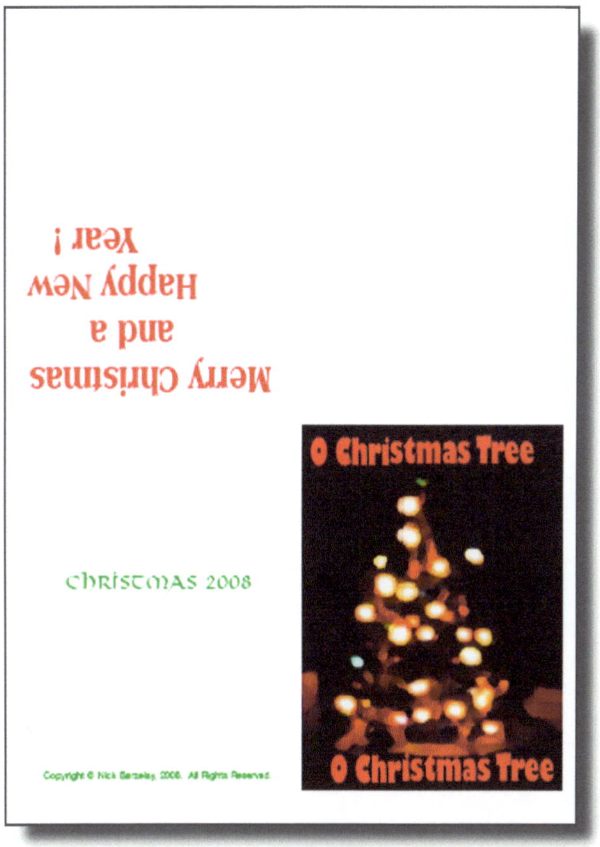

Figure 20-4. Greeting Card Mockup #2

1. What is its purpose of the document; what is it intended to accomplish?

   Obviously, the card is intended to communicate a Christmas greeting. However, because of the potential for content variability, a run of cards could carry any number of greetings and fulfill any number of purposes – for instance a mix of Christmas, New Years, and Hanukkah cards.

2. Who will receive the document and how will it be used?

   Consequently, this second item becomes more significant. The card might be a single document produced on demand for one person. It could be a single design for a specific occasion (Christmas or otherwise) where all recipients gets the same card. Or the card content could vary for each card recipient (each person on the list gets a personalized card of a different type).

3. What content in the document will be static?

   Depending on the previous response, most content could be static (the same content will appear on every card). However, only copyright and series name will be static here. Next question...

4. What content in the document will be variable?

   A mixture of text and imaging could vary in some way. Some text or image may be included on one card but not on another. A related decision is determining the mechanism or data element that will result in the choice of one variable item rather than another one.

5. What data, text, and digital imagery is required for the document?

   Based on our responses so far, we need to conduct inventory to find out just what is available for use, its condition and quality, as well as exactly what additional work will be required to correct any deficiencies. It may be that some content (or all of the content) will have to be developed or acquired externally. Regarding personalization of documents, the data table containing recipients will have to be extended to link each recipient with the appropriate content elements.

6. What will be the dimensions of the document?

   On a more mundane level, this question has probably been answered already, given the mockup or the card template. This is certainly not always going to be the case.

7. What printing capabilities will be available or required to produce the document?

   The question is actually where, how, and with what technology are we going to produce the document. The mockup we are using for this discussion was produced in a home office on an older color inkjet printer capable of only single sided printing – hence the need for folding and upside-down positioning. They could be printed on a color printer (inkjet or laser jet) with duplex print capabilities. Alternatively, the static card or stream of variable cards could be converted to a print quality PDF file and taken to a commercial printer for production on a digital press or color printer.

8. What weight and type of print stock will be needed to produce the document?

   Decisions previously made will play a large part in the selection of stock, but not entirely. The first mockup used a heavier weight, bright white, inkjet-specific paper. The second mockup used gloss-coated, inkjet, photo paper. Final selection depends on what stock a given printer can handle and what artistic affects are desired.

9. What other materials and supplies will be needed for the project?

   The only additional supplies needed for this project were envelopes. Custom envelopes were created for experimental reasons, but commercially produced envelopes in various sizes are available. These could be addressed from the recipient spreadsheet as well.

Possibilities

Given our discussion so far, and for further contemplation, here is a set of wedding stationery (courtesy of my daughter who designed them for her own wedding). They were designed with Adobe Photoshop and InDesign, converted to PDF print-ready format, and taken to a commercial printing establishment for output. They were trimmed, folded, and stuffed by hand for mailing. As presented here, names and places have been changed to protect the innocent or whomever.

None of these documents (Figures 20-5 through 20-10) contain variable content. However, they could be the basis for designing a set of variable wedding stationery. The most obvious approach would be to convert each document design into an InDesign template.

Once a basic set of templates was built, the templates would be used to produce any number of different designs containing variable content elements that depend on recipient, groupings of recipients, or some other criteria to specify who will receive specific invitations and accompanying documents.

One possibility for an international or multi-lingual group could involve varying the set of documents according to each recipient's spoken language.

Rather than personalization, the set of templates could be used to simplify and speed up the design of subsequent variations, each set for different occasions. You can explore more ideas on your own. Here and on succeeding pages is the set of stationary to stimulate your creativity:

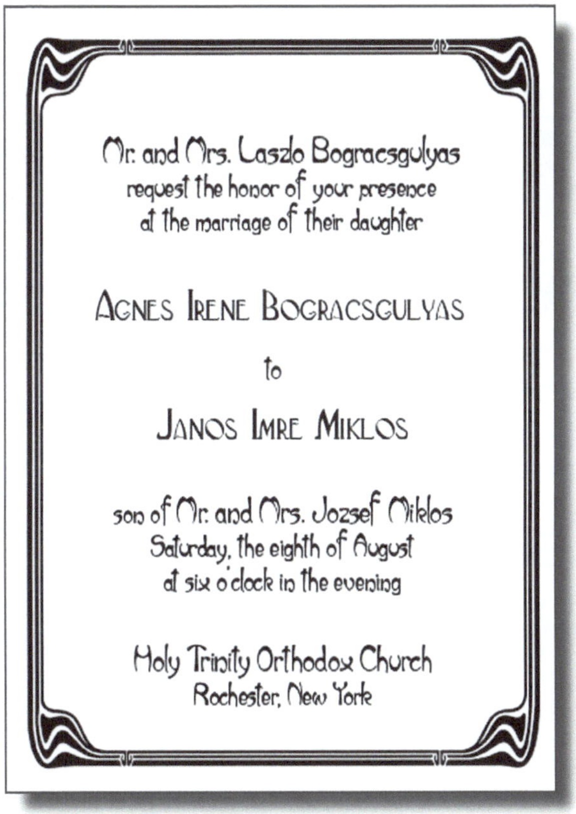

Figure 20-5. Wedding Invitation

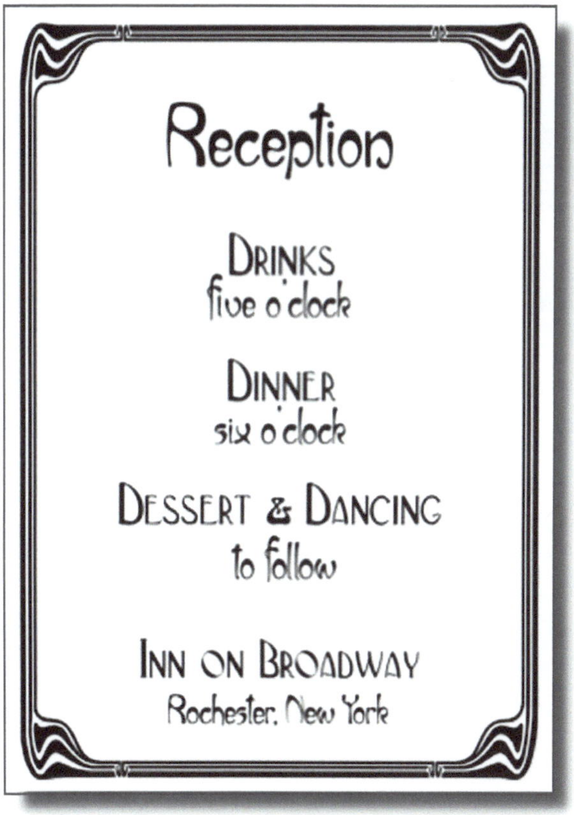

Figure 20-6. Reception Invitation

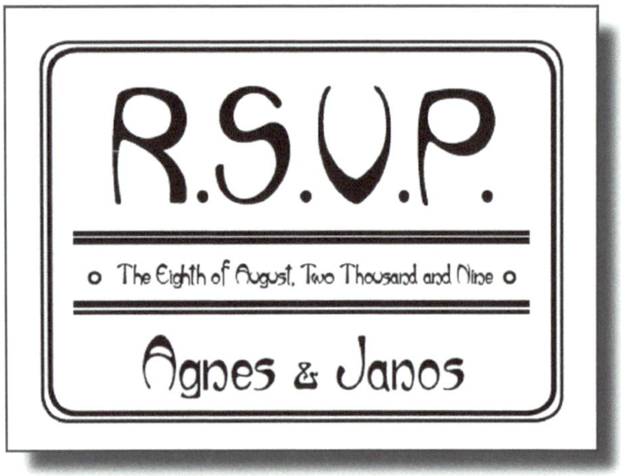

Figure 20-7. RSVP Card (Front)

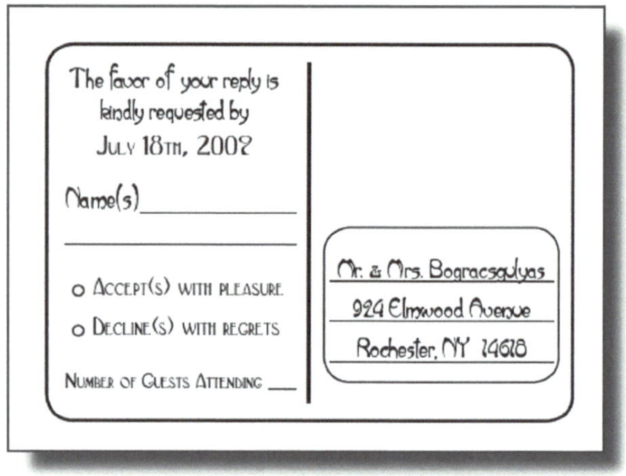

Figure 20-8. RSVP Card (Back)

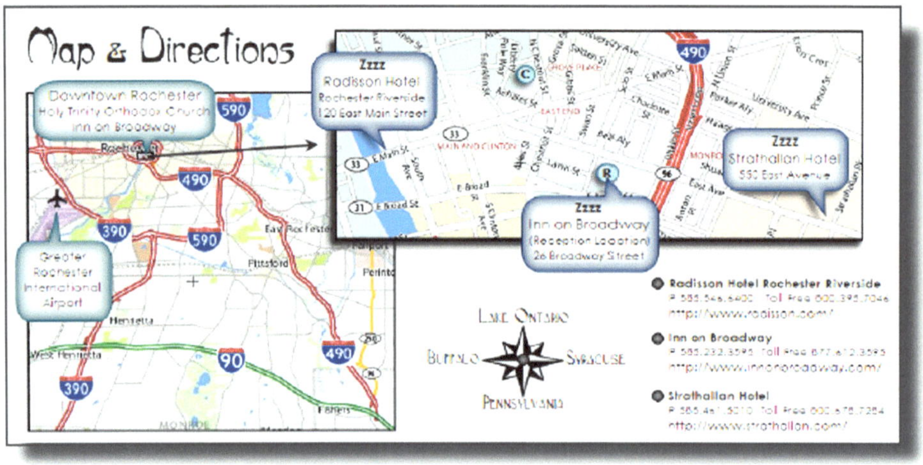

Figure 20-9. Map & Directions Card (Front)

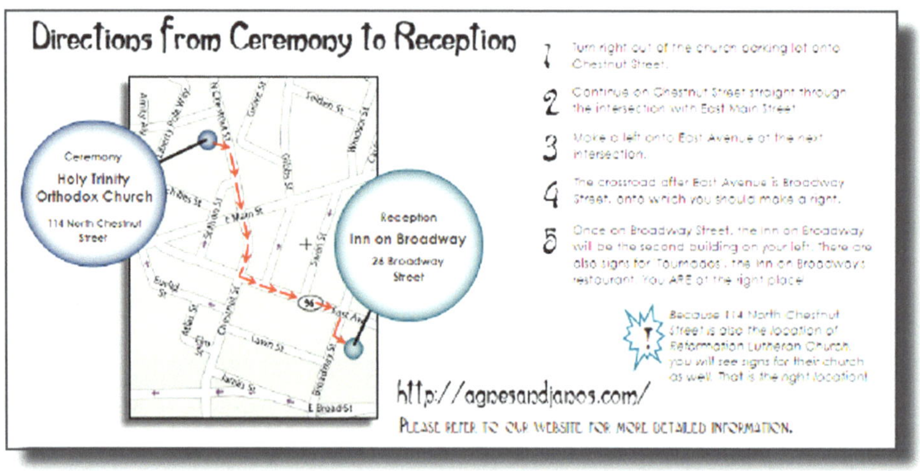

Figure 20-10. Map & Directions Card (Back)

21. A Publishing Encore

Discussion Points:

- ✓ Implication
- ✓ On Reflection
- ✓ Who & Why?
- ✓ Printing Issues
- ✓ Ink On What?
- ✓ Intimation

Implication

Before launching into the last chapter of this book, let me explain a couple things. Most books finish with a concluding chapter; this one doesn't. We've just completed six chapters of demonstrations that were supposed to be thought provoking. This chapter is intended to provoke more thought – this time about putting the results of our work into practice.

Regarding desktop publishing and much larger scales of print production, it is not the intention of this chapter to imply that the tools and techniques presented throughout are only applicable to the very smallest of print jobs. Rather, the idea is to suggest that variable data and content publishing pertain to a whole range of projects: small projects (usually relegated to static techniques), custom or specialty projects, and projects requiring large scale production technology – digital press for variable production runs, and offset press for variable between production runs.

That means variable content publishing or printing can be in everyone's arsenal – the individual, the non-profit, small businesses, large enterprises, and public organizations of every size, digital printers, and offset printers. As for drawing conclusions about the material presented here, each reader will have to rely on their own circumstances and judgement. With that in mind, let's proceed.

On Reflection

Digital publishing has come a long way from the desktop. Much research and development has been aimed at bigger, faster, and better. Market demand is changing – multi-media, high quality, fast turnaround, and lower cost. Competition is hot, driven by a combination of globalization, small fish eaten by large ones, and businesses of all sizes fading out of the market. It's not limited to just printing.

Industry efforts in general appear to continue focusing on speed and quantity with hardware and software geared toward reaching the masses. Tactical shifts using more extensive personalization are avoided with excuses like: "the majority of the market is satisfied with doing mail merge," or "highly skilled programmers are required for more sophisticated VDP but they either aren't available, cost too much, or don't find it interesting."

The impression given is that the "small side" of the market – small and medium sized business – can settle for static documents or mail merge at best. Big business is where it's at, and that translates into big, fast, and expensive. After all, a lot of money can change hands at the top end of the market.

However, embedded within the dynamics of the marketplace is the already sizable and growing importance of small and emerging businesses offering specialized products and services based on innovation and customization. This suggests a growing dichotomy between predominant types of manufacturing or supply activities and new but disregarded sectors of demand.

Who & Why?

Let's refer back to Chapter 1 and the "small and big" discussion. An industry analyst remarked that the VDP approach we have examined is "OK for simple training, but it can't produce a million documents." I understand why a large bank wants to produce a million credit card statements. But what if someone doesn't need to produce a million of something? What if someone doesn't want to communicate with millions, but only reach out to 500 or a few thousand? Consider these:

- A local service business such as lawn care or pool service with a steady customer base
- A specialty retailer catering to repeat business – perhaps a florist, jeweler, or hobby store
- A manufacturer whose customers are either regional wholesalers or brokers
- A low volume custom manufacturer of high cost products for specialized customers
- A local non-profit such as an animal rescue, a performing arts organization, or a political campaign
- A high end, high price, high competition business that needs to make frequent customer contacts to maintain sales volume and customer loyalty – for instance, automobile sales or leasing

Personalized communications using VCP my be extremely effective for any of these examples. What alternatives offer the effectiveness of VCP for low volumes at an affordable cost? One option is the tools and techniques we've discussed, together with some combination of desktop and commercial printing. Here are some possibilities:

1. Design it yourself and print it yourself (if you have the time and also choose to acquire the tools and skills)
2. Design it yourself and go to a "quick" printer to run your print stream – just be careful!
3. Contract with a consultant or designer to do all aspects of the project – especially if it's your first VCP project, but make sure all the "i"s are dotted and the "t"s crossed (some have the right skill set – some don't)
4. Get someone already in printing or publishing (some may only do static work and lack the skills to do VCP)
5. Employ a marketing company (if they can provide end-to-end services at a reasonable price and have the necessary skills to do the job – some may, others may not)
6. Employ a printing company (if they can provide end-to-end services at a reasonable price and have the necessary skills to do the job – some may not have variable-related skills)
7. Use a "quick" printer to do the whole project (if you can find one that has all the skills to get the job done – good luck!)

Whatever choice, careful investigation and judgement will be required to avoid falling on your face or being taken to the proverbial cleaner.

Some projects are going to require the capabilities of a digital press; others may not. Perhaps, a desktop color laser printer may be just the thing. In other cases, a desktop inkjet may offer attractive possibilitie for those who see potential for producing low volume highly customized variable content documents.

The following two topics summarize information and issues that need consideration. In that respect, an article entitled "Printing on Unusual Surfaces" by Rick LePage in the October, 2007, issue of "Macworld Magazine" is very thought-provoking. His article was particularly helpful regarding printers, inks, and substrates. What you will see in the following discussions is an amalgamation of information and thoughts ranging from personal experience, research, advertisements, manufacturing product specifications, and suggestions from LePage's article. First, let's look at likely printing issues when selecting printing technology. Then, based on the technology selection, what appropriate substrates are available and where can they be obtained?

Printing Issues

In selecting printer technology, three dependent factors are in play: printer specifications, ink type, and substrate characteristics – what combination will give the best results for a given set of design requirements. Of course there's always the issue of balancing quality, speed, and cost. While our discussion is slanted toward ink-jet, the issues itemized below may have equal relevance for toner-based electrostatic (laser) printing:

- Printer Use:
  - What is the primary use intended by the manufacturer?
  - Are there other uses this printer can reasonably fulfill?
  - What types of printing should not rely on this printer?
  - Can the printer be shared on a network?
  - How does the printer connect to the computer, and are connections provided?
  - Is additional hardware or software required?
- Printing:
  - What is the color gamut (optimal range of colors) for this printer?
  - Is color selection limited to the standard printing colors (CMYK), or are additional colors (red, blue, orange, or green) available to enlarge the color gamut?
  - Is the printer capable of putting ink all the way to the substrate's edges?
  - Does the printer leave a blank border around the print area, and if so, how wide is it on each side?
  - What is the resolution of the printer?
  - Is the printer capable of duplex printing?
  - If duplex printing, does the printer correctly position (center) the print areas on both sides?
  - Is the printer capable of draft, normal, and high resolution printing, and if so, how does each look and how fast does each setting print?
- Print driver:
  - Is a print driver available for both PCs and Macs?
  - Is the print driver compatible with current or planned operating systems?
  - What printing features and print management settings does the print driver provide?
  - Are there any associated printer utilities along with the driver?

- Ink:
  - Is the ink water-based, solvent-based, or is it toner (remember, we could look at a laser printer)?
  - Is ink color based on pigment (color particles lie on the surface of the substrate, making them susceptible to damage) or dye (coloration is absorbed into the substrate, potentially causing some feathering)?
  - If toner, how scratch resistant is the print and will it require an additional process such as a clear coat?
  - How is the ink packaged, how long does it last, how many sheets can it print (depending on coverage demand), and what is the replacement cost?
- Substrate handling capabilities:
  - How much stock can be loaded into the printer for automatic feed at any one time?
  - Is the option for manual feed (one sheet at a time by hand) available?
  - What are the largest stock dimensions (length and width) the printer is capable of handling?
  - What are the smallest stock dimensions (length and width) the printer is capable of handling?
  - What is the heaviest basis weight stock (pounds for text, cover, and other) that the printer can handle?
  - What the lightest basis weight stock (pounds for text, cover, and other) that a printer can handle?

Beyond how the printer lays ink (or toner) and how the ink (or toner) physically interacts with the substrate, how true to expected color is the printed result. According to LePage (2007), when using specialized substrates, more compelling or satisfactory results may be achieved using the substrate's ICC Color Profile provided by the substrate manufacturer rather than the printer's color profile supplied by the printer manufacturer. Loading the new color profile is a simple three-step process (at least for Mac OS X; PC Windows is probably different):

1. Download the appropriate color profile for the substrate from the manufacturer
2. Place the profile in /Library/ColorSync/Profiles
3. If using Adobe software, turn off the printer color management setting in the Print Dialog Box (This may also be appropriate for other software that will be printing to the specialized stock.)

Ink on What?

Considering what is traditionally termed, "ink on paper", the characteristics of the chosen substrate play an important role. (There's nothing like stating the obvious!) LePage points to these key factors:

- What is the substrate weight (basis: text, cover, other) – 22 to 32 pounds per 500 sheets, or something else?
- What is the substrate brightness (how "white" is it) – US Scale 92 to 100?
- Is the substrate coated on one side, coated on both sides, or not coated at all?
- What is the composition of the substrate?
- Given the dimensions of the substrate (length and width), in which direction does the grain run? (This last is especially critical to prevent toner from cracking when folded, as well as make creasing and folding easier.)

Let's consider how a substrate handles ink to determine what combination will return the best results. This is particularly important when using specialty papers discussed in the LePage article. We'll concentrate on inkjet here. Coloring from pigment-based ink will sit on top of the paper, rather than absorbed into the paper. According to LePage, the following substrates will render good results with pigment:

- Matte paper
- Paper made from fiber
- Fine art paper
- Paper made of fabrics
- Handmade paper

Conversely, coloring from dye-based ink is absorbed by the paper. LePage says the advantage is ink with a potentially wider color gamut; the disadvantage is potential feathering. For good results, we'll want to select papers specifically made for the type of printer we're using and its requirements. Many varieties and brands are available in local stores. These include paper specifically for inkjet or laser printers, photo glossy paper, recycled paper (some may have quality issues), acid-free paper that is "archive safe", bond, and others. For specialty paper, here are some online sources:

- Resin Coated: www.redrivercatalog.com and www.moabpaper.com
- Matte Finish such as "Ultra Premium Presentation Matte" at www.epson.com
- Fiber Based: www.crane.com/museo and www.harman-inkjet.com
- Fine Art: "Photo Rag" at www.hahnemuehle.com
- Fabric: www.colortextiles.com
- Handmade: www.hiromipaper.com

Intimation

As in many other instances, potentially interesting possibilities are generally accompanied by many opportunities, many risks, and many considerations – real and imagined. Each circumstance is likely to be a unique combination of factors that account for who, what, where, when, why, and how. So, as you can probably see, concluding this chapter and this book to everyones satisfaction – tying up all the loose ends and coming up with some proverbial axiom or final word of wisdom – is actually quite an impossibility. You might wonder why. The reason is really quite simple:

Ironically, too much variability!

SIX: INFORMATION

This unit is meant to be informative and helpful – some information you'd expect to find , plus some additional programming assistance in dealing with your XML streams.

Reference

The reference section accounts for all the literature and source material used during research and development for this book.

About the Author

This section makes a few brief comments to acquaint you with the author..

Appendix A – Abbreviations & Acronyms

Appendix A is a list of common technical abbreviations and acronyms used in this book.

Appendix B – Perl Content Stream Utilities

Appendix B contains annotated Perl program code for utilities that provide additional edit capabilities for the XML content stream.

a. Overview
b. About Use
c. Convert
d. Insert
e. Delete
f. Change
g. Reorder
h. Combine
i. Merge
j. Split

Reference

Adobe. (2005). *Adobe InDesign CS2 classroom in a book.* Berkeley, CA: Peachpit Press.

Amiano, M., D'Cruz, C., Ethier, K., & Thomas, M. (2006). *XML problem-design-solution.* Indianapolis, IN: Wiley Publishing, Inc.

Baxes, G. (1994). *Digital image processing principles and applications.* NY: John Wiley & Sons, Inc.

Bishop, B. (2008). *The big sort: the clustering of like-minded America is tearing us apart.* NY: Houghton Mifflin Company.

Boiko, B. (2005). *Content management bible, 2nd edition.* Indianapolis, IA: Wiley Publishing, Inc.

Coffey, G. & Prosser, S. (2005). *FileMaker Pro: the missing manual.* Sebastopol, CA: O'Reilly Media, Inc.

Cooper, P. (2007). *Beginning ruby from novice to professional.* Berkeley, CA: Apress.

Eckes, G. (2003). *Six sigma for everyone.* Hoboken, NJ: John Wiley & Sons, Inc.

Fitzgerald, M. (2007). *Learning ruby.* Sebastopol, CA: O'Reilly Media, Inc.

Fraser, B., Murphy, C., & Bunting, F. (2005). *Real world color management, second edition.* Berkeley, CA: Peachpit Press.

Frey, F. & Barzelay, N. (2008). *Upstream database and digital asset management in variable data printing.* A research monograph of the Print Industry Center. Rochester, NY: Rochester Institute of Technology.

George, M., Rowlands, D., Kastle, B. (2004). *What is lean six sigma?* New York: McGraw-Hill.

Godin, Seth. (2006). *Small is the new big.* NY: Penguin Group (USA), Inc.

Graham, I. (1998). *Requirements engineering and rapid development: an object-oriented approach.* Harlow, England: Addison Wesley Longman, Limited.

Kofler, M. (2005). *The definitive guide to MySQL 5, third edition.* Berkeley, CA: Apress.

LePage, R. (October, 2007). Printing on unusual surfaces – from fine-art paper to fabric, there's a world of possibilities. *Macworld Magazine.*

Lewis, J.P. (1991). *Project planning, scheduling & control – a hands-on guide to bringing projects in on time and on budget.* Chicago: Probus Publishing Company.

Maivald, J. & Palmer, C. (2008). *A designer's guide to Adobe InDesign and XML.* Berkeley, CA: Adobe Press, Peachpit.

Mattison, R. (1999). *Web warehousing and knowledge management.* NY: McGraw Hill.

Meyer, E. A. (2007). *CSS: the definitive guide, third edition.* Sebastopol, CA: O'Reilly Media, Inc.

Paulk, M., Weber, C., Curtis, B., Chrissis, M., et al. (1994). *The capability maturity model: guidelines for improving the software process.* Software Engineering Institute, Carnegie Mellon University. Reading, MA: Addison Wesley Longman, Inc.

Pausch, R. (2007). *The last lecture.* Retrieved September 6, 2008, from WXXI television, Rochester, NY.

Powell, G. (2008). *Instant InDesign: designing templates for fast and efficient page layout.* Berkeley, CA: Adobe Press.

Quigley, E. (2008). *Perl by example, fourth edition.* Boston, MA: Pearson Education, Inc. and Prentice Hall.

Roman, S. (1997). *Access database design & programming.* Sebastopol, CA: O'Reilly & Associates, Inc.

Romano, F., et al. (2002). *Designing for digital.* Alexandria, VA: Digital Printing Council, Printing Industries of America.

Romano, F. (2010). VDP for you and me . . . finally. *WhatTheyThink: Commentary & Analysis.* Retrieved April 2, 2010, from http://whattheythink.com/news/printing-industry-news.cfm

Rud, O. (2001). *Data mining cookbook: modeling data for marketing, risk, and customer relationship management.* NY: John Wiley & Sons, Inc.

Russ, J. (2002). *The image processing handbook, fourth edition.* Boca Raton, FL: CRC Press LLC.

Schmitt, C. (2004). *CSS cookbook.* Sebastopol, CA: O'Reilly Media, Inc.

Schwartz, R., Phoenix, T., & Foy, B. (2005). *Learning Perl, fourth edition.* Sebastopol, CA: O'Reilly Media, Inc.

Stanek, W. (2002). *XML pocket consultant.* Redmond, WA: Microsoft Press.

Taylor, D. (2005). *Learning Unix for Mac OS X Tiger*. Sebastopol, CA: O'Reilly Media, Inc.

Wagner, R. & Mansfield, R. (2003). *XML all-in-one desk reference for dummies.* NY: Wiley Publishing, Inc.

Westphal, C. & Blaxton, T. (1998). *Data mining solutions: methods and tools for solving real–world problems.* NY: John Wiley & Sons, Inc.

About the Author

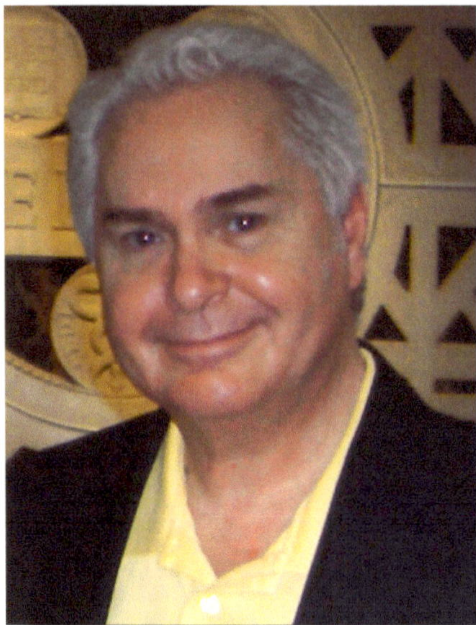

Forget meaningless titles and job descriptions; when all is said and done, I describe myself as a systems analyst and project manager. with many diverse interests and experiences. I've made a career over 30 years in applied information technology and IT project management I earned an M.S. in Print Media from Rochester Institute of Technology and an M.P.A. (Public Administration) from Texas Christian University. My B.A. is in Political Science from the University of Colorado at Boulder.

I've shared responsibility for team teaching cross media workflow and presenting classroom lectures on variable data (or content) printing at Rochester Institute of Technology's School of Print Media. I'm currently determining the best way to release this title as an eBook. I'm also contemplating the idea of publishing some non-technical material using techniques I've been writing about here.

My previous writing is primarily business-related, such as project proposals and analytical papers. More substantial works include:

- A series of blogs for InfoTrends' "The Digital Nirvana": Insights on various aspects of variable data printing.
- "Upstream Database and Digital Asset Management in Variable Data Printing" (co-authored with Dr. Franziska Frey, Ph.D.). R.I.T. Print Industry Center Research Monograph (PICRM200801).
- Project management guide for Xerox Corporate Strategic Services
- Project management and application development methodology for Lawyers Cooperative Publishing Company and Thomson Professional Publishing.

You can learn more about my credentials on LinkedIn (http://www.linkedin.com/in/nickbarzelay). You may contact me through LinkedIn. Please state the nature of your interest and how I can reply. I'll get back to you as soon as I possibly can.

Appendix A – Abbreviations & Acronyms

BOM	Bill of Materials
CMYK	Standard 4-color printing (cyan, magenta, yellow, & black)
CPM	Critical Path Method
CRLF	Carriage Return Line Feed (also used separately – CR or LF)
CRM	Customer Relationship Management
CS	Adobe Creative Suite
CSS	Cascading Style Sheet
CSV	Comma Separated Values
CtP	Computer to Plate
DAM	Digital Asset Management
DAMS	Digital Asset Management System
DB	Database
DBMS	Database Management System
DHTML	Dynamic Hypertext Markup Language
DICOM	Digital Imaging and Communications in Medicine
DOM	Document Object Model
DTD	Document Type Definition
ERP	Enterprise Resource Planning
GPS	Global Positioning System
GUI	Graphical User Interface
HTML	Hypertext Markup Language
ICC	International Color Consortium
IPTC	International Press Telecommunications Council
IRG	Inter Record Gap
IT	Information Technology
LCD	Lowest Common Denominator
NPV	Net Present Value
OR	Operations Research

PDF	Adobe's Portable Document Format (or File)
Perl	Practical Extraction and Report Language
PERT	Program Evaluation & Review Technique
PPBS	Planning, Programming, & Budgeting Systems
QA	Quality Assurance
RGB	Basic digital colors (red, green, & blue)
RIP	Raster Image Processor(or Processing)
RIT	Rochester Institute of Technology
ROI	Return On Investment
SGML	Standard Generalized Markup Language
SPM	School of Print Media
SQL	Structured Query Language
TOC	Table of Contents
URI	Uniform Resource Identifier
URL	Uniform Resource Locator
UTF	Unicode Transformation Format
VBA	Microsoft's Visual Basic for Applications
VCP	Variable Content Publishing (contains subtleties beyond VDP)
VDP	Variable Data Printing (frequently used for both VDP and VCP)
WBS	Work Breakdown Structure
WYSIWYG	What You See Is What You Get
XHTML	Extensible Hypertext Markup Language
XML	Extensible Markup Language
XSL	Extensible Stylesheet Language
XSL-FO	Extensible Stylesheet Language – Formatting Objects
XSLT	Extensible Stylesheet Language Transform

Appendix B – Content Stream Utilities

Discussion & Programs:

- ✓ Overview
- ✓ About Use
- ✓ Convert
- ✓ Insert
- ✓ Delete
- ✓ Change
- ✓ Reorder
- ✓ Combine
- ✓ Merge
- ✓ Split

Overview

This appendix presents a series of programs to generate or manipulate an XML content stream without having to use desktop DBMS facilities. Going all the way back to the DBMS to make revisions there results in re-exporting the data to XML, and ultimately doing another content stream cleanup.

Additionally, the small Perl programs presented here demonstrate some of the possibilities for extending the usefulness of generated XML content streams beyond original intent – making content stream repairs or adjustments to more closely match current document design or support a new document. These programs bypass the need to learn XSLT. They provide example logic for modifying both XML element content and XML hierarchical structure.

Their change capability spans the entire XML stream, while providing the ability to apply modifications selectively. This means the entire content stream can be examined, but only specific targets affected. Targeting can look for specific XML elements, for specific XML elements with particular content, or for elements or content based on some previously read XML element. In other words, everything is accessible and changeable.

The first program takes a delimited file and converts it into an XML content stream that Adobe InDesign can use. Other programs address inserting, deleting, changing, reordering, and combining XML elements. The last two programs deal with merging and splitting XML content streams. Non of these programs use native XML manipulation capabilities such as DOM and XSLT. All operations are performed as if the XML stream was simply another sequential text file.

XML devotees may poke fun or disparage this approach, but it does put XML processing capabilities in the hands of designers, programming neophytes, and others who may not know much about XML beyond tagging. Of course, the assumption we're making is that people unfamiliar with Perl programming (or programming in general) can pick up enough from the program examples to successfully apply them. Why is that?

Because the program logic is already built and only requires some setup information. Points where setup information changes are required will be highlighted in yellow, like this. **Comments within the program that identify individual functions are bold character type, such as this.**

As a refresher, we are referring to the XML expression of a database table or a spreadsheet, which is organized in rows and columns that equate to records and fields. For example, below is an XML row element from the "dog" database. Do you recall that? The tags "<row> and </row>" demarcate the start and end of a single record. The imbedded column elements represent record fields:

```
<row>
    <number>
    <name>
    <breed>
    <gender>
    <photo>
    <weight>
    <age>
</row>
```

About Use

These programs have been tested with sample data. They work. If you decide to use them, be warned that attention to detail is extremely important. After copying, proof your copy. When setting up a program to run with one of your own files, back up your original program copy before making setting changes. Back up your data file as well. A few extra precautions, but you won't be sorry – especially if you're new to Perl, or XML, or programming in general.

There are not a lot of settings to worry about, and most are repetitive. You may see the same or similar setup requirements in multiple programs. Text string settings are very nuanced. For instance, the first character in a string is at position zero, so a ten character string is going to run from position zero through nine. If you are serious about this approach, don't just settle for learning by trail and error. Get yourself an introductory book on this programming language. Books listed in the reference are very helpful.

Convert

There may be a case where data content is provided in a spread sheet or some type of delimited file. For whatever reason, using a desktop DBMS may not be feasible. So how do you convert the data into XML data stream format? If not already in a spreadsheet, import the data into one and then export it back out as a tab-delimited file. If the data is already delimited using something other than tabs, simple search and replace might be an option to replace the original delimiters with tabs. However, the spreadsheet import-export sequence is probably easier.

Why tabs? Because they are not likely to cause any confusion with punctuation imbedded in the content that happens to be the same as the delimiter – commas for instance.

This program provides the capability to take tab-delimited data files (or for that matter, any other kind of delimiter by simply changing the "string split" expression) and convert it into an XML tagged file ready for use with Adobe InDesign. The program also includes the ability to convert any XML element containing image reference data into one with a file reference (href) attribute. There are two caveats:

1. The absolute necessity of understanding the source data file or spreadsheet
2. The absolute necessity of adhering to any conventions established by the program for setup and processing:
   - Set the heading flag (hdflag) to "1" if the first data row contains headings; otherwise, leave it at "0" (zero).
   - If the file doesn't have headings, change the program's default tag names to match the data columns (same number of columns in the same order).
   - Insure that "line feeds" at the end of each row in the delimited file conform to operating system requirements. Mac OS X will use a standard UNIX "new line" (\n). Windows operating systems will specify something else. The program will not perform properly without the correct symbol.
   - Elements containing external file or image references should adhere to the format, "file://URI path/file name with extension". The program explicitly searches for "file://" to identify content requiring conversion to "href" attributes. Here's an example from the "Dog Database": file://desktop/images/slinky.jpg".

Convert Tab-Delimited Files

```
#!/usr/bin/perl
use warnings;
use strict "subs";

# Set input and output file names
$inputfl = "DataFileIn.tab";
$outputfl = "XMLFileOut.xml";
$errlist = "ErrListOut.txt";

# Static variables
$declaration = '<?xml version="1.0" encoding="UTF-8"?>';
$rootfore = "<dataroot>\n";
$rootaft = "</dataroot>\n";
$rowfore = "<row>\n";
$rowaft = "</row>\n";
$tagfore = "<";
$tagaft = ">";
$tagterm = "</";
$tagend = ">\n";
$href = " href=\"";
$hrefend = "\"/>\n";
$reftest = "file://";

# XML element processing preparations

# Column names can be revised and increased or decreased
@tagName = qw/col1 col2 col3 col4 col5 col6 col7/;
$cols = @tagName;

# Set flag to 1 if first line of input contains column headings
$hdflag = 0;
if (not $hdflag) {&prepTags;}

# Conversion processing
&openFiles($inputfl, $outputfl, $errlist);
$export = $declaration . "\n" . $rootfore;
print OUTFILE $export;
$export = "DELIMITED DATA ERROR LISTING\n\n";
print ERRFILE $export;
```

```
$errcount = 0;
&convert2xml;
```

Read delimited data in, write XML out

```
sub convert2xml {
        while ($import = <INFILE>) {
            chomp($import);
            @rowBuilder = split("\t", $import);
            $check = @rowBuilder;
            if ($hdflag) {
                $hdflag = 0;
                @tagName = @rowBuilder;
                $cols = @tagName;
                &prepTags;
                next;
            }
            if ($check != $cols) {
                $errcount++;
                $export = "Item Error " . $errcount . ": " . $import . "\n";
                print ERRFILE $export;
                next;
            }
            $export = &tagContent;
            print OUTFILE $export;
        }
        $export = $rootaft;
        print OUTFILE $export;
        if ($errcount == 0) {
            $export = "No data errors found!\n";
            print ERRFILE $export;
        }
        &closeFiles;
}
```

Prepare XML element tags

```
sub prepTags {
        @tagFront = @tagName;
        @tagBack = @tagName;
        foreach $tag (@tagFront) {
            $tag = $tagfore . $tag . $tagaft;
        }
        foreach $tag (@tagBack) {
            $tag = $tagterm . $tag . $tagend;
        }
}
```

Apply XML tags to row content

```
sub tagContent {
        for ($i = 0; $i < $cols; $i++) {
            if (substr($rowBuilder[$i], 0, 7) eq $reftest) {
                $tag = substr($tagFront[$i], 0, (length($tagFront[$i]) - 1));
                $rowBuilder[$i] = $tag . $href . $rowBuilder[$i] . $hrefend;
        }
        else {
            $rowBuilder[$i] = $tagFront[$i] . $rowBuilder[$i] . $tagBack[$i];
        }
        }
        $newXMLrow = $rowfore . join("",@rowBuilder) . $rowaft;
```

```
                return $newXMLrow;
}

# File open process
sub openFiles {
                $infl = $_[0];
                $outfl = $_[1];
                $errfl = $_[2];
                open(INFILE, "<$infl") || die "Can't find or open $infl: $!";
                open(OUTFILE, ">$outfl") || die "Can't open $outfl: $!";
                open(ERRFILE, ">$errfl") || die "Can't open $errfl: $!";
}

# File close process
sub closeFiles {
                close(INFILE);
                close(OUTFILE);
                close(ERRFILE);
}
```

Listing B-1. Convert Tab-delimited Files to XML Content Streams

INSERT

This program inserts one or more XML elements at a specified element location within each <row> element of the XML content stream.

Insert New XML Elements

```
#!/usr/bin/perl
use warnings;
use strict "subs";

# set input and output
$inputfl = "InputName.xml";
$outputfl = "OutputName.xml";
&insertProcess;

# Open file process
sub openFiles {
                $infl = $_[0];
                $outfl = $_[1];
                open(INFILE, "<$infl") || die "Can't find or open $infl: $!";
                open(OUTFILE, ">$outfl") || die "Can't open $outfl: $!";
}

# Close file process
sub closeFiles {
                close(INFILE);
                close(OUTFILE);
```

```
}

# Insertion processing
sub insertProcess {
            &openFiles($inputfl, $outputfl);
            $addFlag = 0;

    # Number of new elements
            $new = 3;

    # List of new elements
            @newElements = ("<one>Insert First Element</one>",
                    "<two>Insert Second Element</two>", "<three>Insert third element</three>");

    # File processing main loop
            while ($import = <INFILE>) {
                chomp($import);
                if ($addFlag) {
                    for ($i = 0; $i < $new; $i++) {
                        $export = $newElements[$i] . "\n";
                        print OUTFILE  $export;
                    }
                    $addFlag = 0;
                }
                else {
                    $export = ($import . "\n");
                    print OUTFILE $export;
            # Tag location after which new elements are inserted
            # Test string length and name
                    if (substr($import,0,7) eq "<offer>") {$addFlag = 1;}
                }
            }
            &closeFiles;
}
```

Listing B-2. Insert New XML Elements Into the Content Stream

Delete

This program deletes one or more XML elements from each <row> of the XML content stream.

Delete XML Elements

```
#!/usr/bin/perl
use warnings;
use strict "subs";

&deleteProcessing;
```

```
# set input and output
$inputfl = "InputName.xml";
$outputfl = "OutputName.xml";

# Open file process
sub openFiles {
        $infl = $_[0];
        $outfl = $_[1];
        open(INFILE, "<$infl") || die "Can't find or open $infl: $!";
        open(OUTFILE, ">$outfl") || die "Can't open $outfl: $!";
}

# Close file process
sub closeFiles {
        close(INFILE);
        close(OUTFILE);
}

# Delete process
sub deleteProcessing {
        &openFiles($inputfl, $outputfl);
        while ($import = <INFILE>) {
            chomp($import);

    # Tag length and name to be deleted
            if (substr($import,0,4) eq "<one") {next;}
            if (substr($import,0,4) eq "<two") {next;}
            if (substr($import,0,6) eq "<three") {next;}
            $export = $import . "\n";
            print OUTFILE $export;
        }
        &closeFiles;
}
```

Listing B-3. Delete XML Elements From the Content Stream

CHANGE

This program completely changes the content of one or more selected elements from each <row> element of the XML content stream.

Change XML Elements

```
#!/usr/bin/perl
use warnings;
use strict "subs";

# set input and output
$inputfl = "InputName.xml";
$outputfl = "OutputName.xml";
```

```
&changeProcessing;

# Open file process
sub openFiles {
        $infl = $_[0];
        $outfl = $_[1];
        open(INFILE, "<$infl") || die "Can't find or open $infl: $!";
        open(OUTFILE, ">$outfl") || die "Can't open $outfl: $!";
}

# Close file process
sub closeFiles {
        close(INFILE);
        close(OUTFILE);
}

# Change process
sub changeProcessing {
        &openFiles($inputfl, $outputfl);
        while ($import = <INFILE>) {
            chomp($import);

    # Set change tag test length and name
            if (substr($import,0,10) eq "<location>") {

            # Set decision value position and length
                $loc = substr($import,10,1);
            }

    # Set test length, tag name, and replacement values
            if (substr($import,0,7) eq "<offer>") {
                if ($loc eq "N") {$export = ("<offer>5% Discount</offer>" . "\n");}
                if ($loc eq "S") {$export = ("<offer>10% Discount</offer>" . "\n");}
                if ($loc eq "E") {$export = ("<offer>15% Discount</offer>" . "\n");}
                if ($loc eq "W") {$export = ("<offer>20% Discount</offer>" . "\n");}
            }

    # Additional change settings
            elsif (substr($import,0,9) eq "<product>") {
                if ($loc eq "N") {$export = ("<product>Paving Stone</product>" ."\n");}
                if ($loc eq "S") {$export = ("<product>Fertilizer</product>" . "\n");}
                if ($loc eq "E") {$export = ("<product>Garden Tools</product>" ."\n");}
                if ($loc eq "W") {$export = ("<product>Seed Packets</product>" ."\n");}
            }
            else {
                $export = $import . "\n";
            }
           print OUTFILE $export;
        }
        &closeFiles;
}
```

Listing B-4. Change Selected XML Elements in the Content Stream

Reorder

This program reorders (changes the sequence) of each element within each <row> element pair of the XML content stream.

Reorder XML Elements

```
#!/usr/bin/perl
use warnings;
use strict "subs";

# set input and output
$inputfl = "InputName.xml";
$outputfl = "OutputName.xml";

&reorderProcess;

# Open file process
sub openFiles {
        $infl = $_[0];
        $outfl = $_[1];
        open(INFILE, "<$infl") || die "Can't find or open $infl: $!";
        open(OUTFILE, ">$outfl") || die "Can't open $outfl: $!";
}

# Close file process
sub closeFiles {
        close(INFILE);
        close(OUTFILE);
}

# Reorder XML process
sub reorderProcess {
        &openFiles($inputfl, $outputfl);
    # set up operating values
    # desired order of row elements in new file
        @order = (10, 3, 4, 5, 6, 7, 8, 9, 2, 1);

    # iteration limit is number of elements within a row
        $limit = 10;
        # array of elements within a row
        @content = ();
        $flag = "N";  # process control flag
        # iteration counter
        $count = 0;
        while ($import = <INFILE>) {
            chomp($import);
            # test for row content processing
            if ($flag eq "Y") {
                $content[$count] = ($import . "\n");
                # test for end row content completion
                if ($count == $limit) {
                    # row content reorder processing
                    for ($i = 0; $i < $limit; $i++) {
                        $key = ($order[$i] - 1);
```

```
                        $export = $content[$key];
                        print OUTFILE $export;
                    }
                    $flag = "N"; # turn off row flag
                }
                $count++;
            }
            else {
                $export = ($import . "\n");
                print OUTFILE $export;
                # test for row
                if (substr($import,0,5) eq "<row>") {
                    $flag = "Y"; # turn on row flag
                    $count = 0;
                }
            }
        }
        &closeFiles;
}
```

Listing B-5. Reorder XML Child Elements Within Each <row> Set

Combine

Here is some background. A mailing address may have an additional address line (i.e. Building 15 Suite 112, 32 Canal Parkway). How do you compensate for the "structural variability" of the XML content stream? A document design with double addresses (two XML elements) will create a blank line when encountering single address data. A document design with a single address field won't handle double addresses properly. One solution is to use a single address layout, combine double addresses into single XML elements, and let the XML import process handle differences.

Combine Two Consecutive XML Elements

```
#!/usr/bin/perl
use warnings;
use strict "subs";

# set input and output
$inputfl = "InputName.xml";
$outputfl = "OutputName.xml";

&combineElements;

# Open file process
sub openFiles {
        $infl = $_[0];
        $outfl = $_[1];
        open(INFILE, "<$infl") || die "Can't find or open $infl: $!";
        open(OUTFILE, ">$outfl") || die "Can't open $outfl: $!";
}
```

```
# Close file process
sub closeFiles {
            close(INFILE);
            close(OUTFILE);
}

# Element combine process
sub combineElements {
            # Set element tag name patterns
            $pat1 = "<address1>";
            $pat2 = "<address2>";
            $pat3 = "<address2/>";
            $pat4 = "<address2><";
            &openFiles($inputfl, $outputfl);
            # File processing loop
            while ($import = <INFILE>) {
                chomp($import);
                if (substr($import,0,10) eq $pat1) {
                    @tags2a = evalElement($import);
                    next;
                }
                if (substr($import,0,11) eq $pat3 or substr($import,0,11) eq $pat4) {
                    $import = join("",@tags2a); }
                elsif (substr($import,0,10) eq $pat2) {
                    @tags2b = evalElement($import);
                    if ($tags2b[1] =~ m/^[ +|\0]/ ) {
                        $import = join("",@tags2a);}
                    else {
                        $combine = $tags2a[1] . "\n" . $tags2b[1];
                        $tags2a[1] = $combine;
                        $import = join("",@tags2a);
                    }
                }
                $export = $import . "\n";
                print OUTFILE $export;
            }
            &closeFiles;
}

# Element analysis subroutine
sub evalElement($) {
            $element = shift(@_);
            $lentxt = length($element);
            $boundary1 = index($element,">");
            $offset = $boundary1 + 1;
            $boundary2 = ($lentxt - rindex($element,"<"));
            $sublen = $lentxt - $offset - $boundary2;
            $content = substr($element, $offset, $sublen);
            $tag = substr($element, 0, $offset);
            $endtag = substr($element, ($offset + $sublen));
            @elementArray = ($tag, $content, $endtag);
            return @elementArray;
}
```

Listing B-6. Combine Contiguous XML Child Elements Within a <row> Set

Merge

Imagine that we have three XML files (same type, different content). They've been used separately, but now we want to bring them together in a single content stream that satisfies XML language conventions. This program takes one or more XML content streams and appends them together – one after another – as a single "well formed" XML file.

Merge XML Files

```
#!/usr/bin/perl
use warnings;
use strict "subs";

# List files to be merged
# First file name in list is the destination file
# Remaining file names in list are subsequent source files for merger
&mergeProcess("MergeOut.xml", "MergeOne.xml", "MergeTwo.xml", "MergeThree.xml", "MergeFour.xml");

# Merge files process
sub mergeProcess {
        $outfl = shift(@_);
        @mergelist = @_;
        $count = $#mergelist;
        $cycle = 0;
        $proc = "<?xml";
        $root = "<dataroot>";
        $endroot = "</dataroot>";
        open(OUTFILE, ">$outfl") || die "Can't open $outfl: $!";
        foreach $infl (@mergelist) {
            open(INFILE, "<$infl") || die "Can't find or open $infl: $!";
            while ($import = <INFILE>) {
                chomp($import);
                if (substr($import,0,5) eq $proc and $cycle > 0) {next;}
                if (substr($import,0,10) eq $root and $cycle > 0) {next;}
                if (substr($import,0,11) eq $endroot and $cycle < $count) {next;}
                $export = $import . "\n";
                print OUTFILE $export;
            }
            close(INFILE);
            $cycle += 1;
        }
        close(OUTFILE);
}
```

Listing B-7. Merge (Append) Multiple XML Content Streams

Split

This is a reverse of the previous situation. We have a single large XML content stream that we want to divide into two or more files, each having a set number of records, and each adhering to XML conventions – each new file is well formed. That's what this program does. There may likely be other criteria for making the split, rather than record count. That logic can easily be added to this program, replacing the record count logic.

Split XML File Into Smaller Files

```
#!/usr/bin/perl
use warnings;
use strict "subs";

# set input and output
$inputfl = "SplitSource.xml";
$outputfl = "SplitSeg1.xml";

$firstline = '<?xml version="1.0" encoding="UTF-8"?>';
$secndline = "<dataroot>";
$endrecord = "</row>";
$endfile = "</dataroot>";

&splitFile;

# Open file process
sub openFiles {
            $infl = $_[0];
            $outfl = $_[1];
            open(INFILE, "<$infl") || die "Can't find or open $infl: $!";
            open(OUTFILE, ">$outfl") || die "Can't open $outfl: $!";
}

# Close file process
sub closeFiles {
            close(INFILE);
            close(OUTFILE);
}

# Split file process
sub splitFile {
            &openFiles($inputfl, $outputfl);
            &recordCount;

    # Processing variables
    # Maximum umber of records per file
            $rcrdlimit = 13;
            $rcrdcount = 0;
            $rcrdaccum = 0;
            $filecount = 1;
            # File processing loop
            while ($import = <INFILE>) {
                chomp($import);
                if ($import eq $endfile) {last;}
                if (substr($import,0,6) eq $endrecord) {
                    $rcrdcount += 1;
```

```
                $rcrdaccum += 1;
            }
            $export = $import . "\n";
            print OUTFILE $export;
            # Check the split count
            if ($rcrdcount == $rcrdlimit) {
                $rcrdcount = 0;
                $filecount += 1;
                $export = $endfile . "\n";
                print OUTFILE $export;
                close(OUTFILE);
                # Check for last record
                if ($rcrdaccum < $rcrdtotal) {
                    substr($outfl,8,1) =  $filecount;
                    open(OUTFILE, ">$outfl") || die "Can't open $outfl: $!";
                    $export = $firstline . "\n";
                    print OUTFILE $export;
                    $export = $secndline . "\n";
                    print OUTFILE $export;
                }
            }
        }
        &closeFiles;
}

# Count completely processed records
sub recordCount {
        $rcrdtotal = 0;
        while ($import = <INFILE>) {
            chomp($import);
             if (substr($import,0,6) eq $endrecord) { $rcrdtotal += 1;}
        }
        print "Record Count: $rcrdtotal \n";
        open(INFILE, "<$infl") || die "Can't find or open $infl: $!";
}
```

Listing B-8. Split a Single XML Content Stream into Multiple Streams

www.ingramcontent.com/pod-product-compliance
Ingram Content Group UK Ltd.
Pitfield, Milton Keynes, MK11 3LW, UK
UKHW060016051026
15591UKWH00002B/232
9780615482811